Science and Society in
Restoration England

Science and Society in Restoration England

MICHAEL HUNTER

CAMBRIDGE UNIVERSITY PRESS

CAMBRIDGE

LONDON   NEW YORK   NEW ROCHELLE
MELBOURNE   SYDNEY

Published by the Press Syndicate of the University of Cambridge
The Pitt Building, Trumpington Street, Cambridge CB2 1RP
32 East 57th Street, New York, NY 10022, USA
296 Beaconsfield Parade, Middle Park, Melbourne 3206, Australia

First published 1981

Phototypeset in Linotron 202 Sabon by
Western Printing Services Ltd, Bristol
Printed and bound in Great Britain
at The Pitman Press, Bath

British Library Cataloguing in Publication Data

Hunter, Michael
Science and society in Restoration England.
1. Science – Social aspects – England
2. Science – England – History – 17th century
I. Title
301.24'3'0942   Q175.52.G7   80-41071
ISBN 0 521 22866 2 hard covers
ISBN 0 521 29685 4 paperback

For
DAVID

Contents

mnmnmnmnmnmnmnmnmnmnmnmnmnmnmnmnmnmnmnmnmnmnmnmnmnmnmnmnmnmnmnmnmnmnmnmnmnmnmnmnmn

Preface

This book is based not only on research but on much reflection about science and its role in early modern England. It owes a good deal to a loyal discussion group that met at Birkbeck College, London, during 1977 and 1978, where I was able to ventilate and modify my ideas and prejudices. I must also record my thanks to the seminar audiences at various universities on whom I have tried out preliminary versions of the following chapters: chapter 2 at Oxford in 1974; chapter 3 at University College, London, in 1976 and Birmingham in 1977; chapter 5 at Leeds in 1979; and chapter 6 at Oxford and Sussex in 1978. I am grateful to my hosts on those occasions, namely Lord Dacre, Bill Bynum, David Allen, Paul Wood, Keith Thomas and Peter Burke. I have two further debts: one, for the opportunity to compile and publish a lengthy analysis of the Royal Society's early membership in the Society's *Notes and records* in 1976, in which I was encouraged not least by Sir William Paton; the other, for the stimulus provided by J. R. Jones in asking me to contribute to his recent volume in the 'Problems in focus' series. My chapter in *The restored monarchy* is a preliminary summary of the theme of this book, around which my ideas have formed and developed.

Much of the research for this book was done while I held research fellowships successively at Worcester College, Oxford, and in the History Department at the University of Reading: I am thankful to both institutions for the opportunities thus provided. The research was finished and the book written since I have been at Birkbeck, and I must acknowledge the forbearance and encouragement of my colleagues in the History Department.

The following have read all or part of the book and I am grateful for their comments: David Bebbington, Leslie Clarkson, Estelle Cohen, Eamon Duffy, Moti Feingold, Graham Gibbs, Marie Boas Hall, John Henry, Jim Jacob, Fritz Levy, Joseph Levine, Christine

McLeod, John Miller, Tom O'Malley, Charles Schmitt, Joan Thirsk and Anthony Turner. Some of these have also kindly answered specific queries, as have Alan Gabbey, Fred Lock, Mark Overton, Roy Porter, Charles Webster and Christopher Wright; Annabel Gregory has assisted with my research. I owe a special debt to Sheridan Gilley and – above all – to Paul Wood, who has discussed the book with me innumerable times. Even he disagrees with a few of its arguments and other readers have taken issue with more. It is now for a larger audience to decide who is right.

M.C.W.H.

Hackney
January 1980

List of abbreviations

mmmmmmmmmmmmmmmmmmmmmmmmmmmmmmmmmmmmmmmmmmmmmmmmmmmmmmmmmmmmmmmmmmmmmmmm

Add.	British Library Additional Manuscript
Aubrey	Bodleian Library, Oxford, manuscript Aubrey
B.J.H.S.	*British Journal for the History of Science*
Cl. P.	Royal Society Classified Papers
C.S.P.	*Calendar of State Papers*
DM	Royal Society Domestic Manuscripts
EL	Royal Society Early Letters
Evelyn, *Corr.*	Christ Church, Oxford, Evelyn collection: bound volumes of correspondence
Evelyn, *Lb.*	John Evelyn's Letter-book, Christ Church, Oxford, Evelyn collection manuscript 39 (letters in main numerical series)
H.M.C.	*Historical Manuscripts Commission*
Hales, *Account*	[Thomas Hales], *An Account of several New Inventions and Improvements Now necessary for England* (London, 1691)
JBO	Royal Society Original Journal Book
J.H.I.	*Journal of the History of Ideas*
LBO	Royal Society Original Letter Book
Lister	Bodleian Library, Oxford, manuscript Lister
N. & R. R. S.	*Notes and Records of the Royal Society*
Pepysian	Magdalene College, Cambridge, Pepysian Library manuscript
Phil. Trans.	*Philosophical Transactions*
RBO	Royal Society Original Register Book
Rawlinson	Bodleian Library, Oxford, manuscript Rawlinson
Sloane	British Library manuscript Sloane

Note: authors' names followed by raised numerals in the footnotes refer to entries in the bibliographical essay, pp. 198–219.

Introduction

mnmnmnmnmnmnmnmnmnmnmnmnmnmnmnmnmnmnmnmnmnmnmnmnmnmnmnmnmnmnmnmnmnmnmnmnmnmnmnmnmnmnmnmnmnmn

'Since the rise of Christianity, there is no landmark in history that is worthy to be compared with this'.[1] So Sir Herbert Butterfield acclaimed the Scientific Revolution of the seventeenth century, and such high estimates have assured that the science of the time has been subjected to intense and prolonged scrutiny. Not only have scientific ideas themselves been much studied. Attention has also been paid to their setting, since historians have long believed that the phenomenal intellectual fertility of this period must be linked to the social and economic milieu in which it occurred and must in turn itself have had broader effects.

Modern concern with such questions dates largely from the 1930s. Its most influential pioneers were the Russian physicist, Boris Hessen, and the American sociologist, Robert K. Merton, though an honourable place may also be found for the English historian, Sir George Clark, whose *Science and social welfare in the age of Newton* (1937) arguably says as much in its brief format as any work on a related topic since.[2] The views of these scholars differed markedly and the subject has always been characterised by lively controversy. Indeed contrasting viewpoints have been propounded on numerous topics – from the social and ideological forces encouraging creativity to the links between science and technology, the significance of scientific organisations and the acceptance of science and scientific norms in the community at large.

Some have attempted generalisations about science and society in early modern Europe as a whole, but attention has focused mainly

---

[1] H. Butterfield, *The origins of modern science 1300–1800* (London, 1949), p. 174.
[2] Hessen,[167] Merton,[168] Clark.[172] Another early study of a different kind is Jones,[240] first published in 1936. Here and throughout the notes, citations of an author's name with a raised numeral refer to entries in the bibliographical essay, pp. 198–219.

on seventeenth-century England, the subject of Hessen's, Merton's and Clark's essays. This is not entirely surprising. Science in the Stuart era cries out to be linked to other features of what was by any account a formative period of English history: an age of political revolution and religious conflict, of economic development and social change. Many studies have been made of science and of these broader themes, on a larger or smaller scale and with more or less satisfying results, and the profuse relevant literature that now exists is surveyed in the bibliographical essay appended to this work. This is partly a reference guide, but it is also intended to set the scene for the reappraisal of each aspect of science's role put forward in the chapters that precede it.

The terms of reference of this book are limited. 'England' is fairly strictly interpreted throughout: there are some allusions to published work on Ireland and evidence from Wales and Scotland has occasionally been introduced to illustrate specific points, but no attempt has been made at a systematic survey of events in those countries. More significantly, it is confined to the Restoration, to the years between the return of the Stuarts in 1660 and the end of the seventeenth century.

Its relatively narrow focus is deliberate: so important a subject as the role of science in this formative period is much prone to facile oversimplification, and this book tries to do justice to the subtlety and complexity of the questions it raises. A study of Restoration England may be justified as intrinsically valuable. Not only were these the chief productive years of Newton, Boyle, Hooke and a host of other exponents of the 'new' experimental philosophy; they also witnessed the foundation of the first English scientific institution, the Royal Society. But, though primarily applicable to these decisive decades, many of the conclusions drawn here about the setting and impact of the new science have direct relevance to broader issues in England and Europe. They thus throw new light on many of the imponderables about the social dimension of early modern science that have long preoccupied scholars.

Moreover, though the book is mainly about the Restoration, an attempt is made – especially in the first chapter – to relate trends after 1660 to what had gone before. These earlier developments have recently been considered in detail – particularly by Charles Webster in *The Great Instauration* (1975) – and the angle taken here may help to balance this intensive work on the Interregnum. To be truly

convincing, views about science's role and appeal have to account
for its equal success before and after 1660.

The Restoration was an ambivalent era, in which reaction against
the events of the previous two decades was offset by more positive
features: the affiliations of science have to be seen in terms of the
balance, and sometimes the tension, between these. Many features of
the Interregnum were rejected in a manner characteristic of a 'post-
Revolutionary' regime. All legislation passed without royal assent
was expunged from the Statute Book and there was a general reaction in favour of conservatism and royalism and against the possibility of new disorder. A deprecation of Republicanism and its
corollaries was accompanied by a reversal of the liberalising tendencies of the earlier regime in politics, law and society: this was perhaps
symbolised by the newly broad Statute of Treasons of 1661 and the
reimposition of repressive and unpopular licensing laws presided
over by the redoubtable Sir Roger L'Estrange. The general revulsion
against Puritan fanaticism was epitomised by the vast popularity of
Samuel Butler's *Hudibras*. Its most tangible and significant outcome
was the restoration of the Anglican church in a form which, due to
the heat of reaction among Members of Parliament and others, was
more rigid and doctrinaire than the King and other architects of the
Restoration settlement had originally intended.[3] These circumstances favoured a conservative element in the universities which
had survived threats of reform and even extradition in the revolutionary years, while the free thought which the confusion of the
Interregnum was seen to have encouraged was also widely deprecated.

Yet the reaction was not so complete as to re-create the state of
affairs before the Revolution. The Interregnum had seen great
growth in the power and efficacy of government: newly heavy taxes
like the excise had come to be taken for granted, intervention by the
central government in the localities had become commoner, standing armed forces had been retained, and the civil service had grown
in size and efficiency. These features, which Charles II inherited,
added a new element to politics which had been lacking in his
father's time, in the form of a powerful executive and the real
possibility of the development of centralised efficiency in the direction of royal absolutism – a development familiar elsewhere in

---

[3] Whiteman in Nuttall and Chadwick,[10] pp. 19–88; I. M. Green, *The re-establishment
of the Church of England 1660–63* (Oxford, 1978).

seventeenth-century Europe, particularly in France. Impulses towards absolutism, however, coexisted throughout this period with a
chronic political instability which was equally characteristic of the
time, for the Civil War had hardly solved any of the constitutional
issues between King and Parliament over which it had been fought.
Instead, these continued to be debated fiercely and inconclusively,
and were made more intense by fears of the growing power of the
executive. In the resulting tension, scientists ambitious to implement
reforms took up a clear position which will be outlined in chapter 5:
their loyalties were to strong government, and official support for
them, though generally disappointing, reached a climax when royal
power was on the ascendent in the 1680s.

   The Restoration also had an important inheritance from the Interregnum in economic affairs. In contrast to the bleakness of the early
seventeenth century, the English economy was booming in the years
after about 1650. Its strength was due largely to the burgeoning
re-export trade, which laid the foundations of England's more general commercial prosperity in the eighteenth century, but a lesser
contribution came from agricultural improvement and industrial
change. Current scholarship does not on the whole indicate that this
was due to the policy of any particular regime. Attempts were made
both before and after the Restoration to encourage it through the
so-called Navigation Acts, however, and this economic vitality was
greeted by contemporaries with a general patriotic enthusiasm.[4]

   There was clearly an indirect link between science and economic
life, for the prevalent innovative attitudes in science were mirrored in
the fertility of invention shown by patent applications and in a
widespread optimism about the potential for improvement.[5] Indeed
there were hopes in scientific circles that the link might become
direct, with national prosperity and useful, scientific knowledge
advancing hand in hand. Thus it might be possible 'to render England the Glory of the Western World, by making it the Seat of the
best knowledge, as well as it may be the Seat of the greatest Trade',
with merchants patronising and assisting intellectual activity, while
it was also hoped that the findings of the new science could improve
techniques in agriculture and industry.[6]

---

[4] Wilson,[13] pp. 61–4, 163–4 and passim; Jackson in Jones,[2] pp. 153–4.
[5] K. G. Davies, 'Joint-stock investment in the later seventeenth century', *Economic
   History Review*, 2nd series 4 (1952), 283–5.
[6] Oldenburg to Rycaut, 30 Jan. 1668, Oldenburg,[121] IV, 133.

But – as chapters 3 and 4 show – modern scholars have been misled who have taken these hopes of the propagandists of the new science at their face value instead of finding out how far they were realised. The links between science and the mercantile community proved disappointing, and, though the interest of intellectuals in technological improvement is significant, the extent to which they contributed to economic development is questionable. If this has sometimes been recognised – particularly by economic historians – little attempt has hitherto been made to explore the reasons for it; these are examined in chapter 4 and they throw interesting light on the 'cultural topography' of the period.

The presumption that science and technology must have been closely linked has distracted attention from another crucial development that forms the background to science. This was the growth of a fashionable, leisured culture focused on London – the seat of government and law – which had such important consequences as the growth of the retail trade and a general increase in consumption which has sometimes been seen as contributing to the Industrial Revolution.[7] This is the setting of the culture of the 'virtuosi' and of a significant public audience for intellectual matters whose links with science will be considered in chapter 3. It was a culture which grew phenomenally in the late seventeenth century, while so did London social life, epitomised by the rapid spread of coffee houses from their first introduction there in 1652.[8] Though London was the centre of this – as it was the focus of most cultural and economic activity – it also made inroads in the provinces. The relations between London and the country and their significance for science will also be surveyed in chapter 3, where some attempt will be made to assess the receptiveness of English society to the new science at all levels.

Such social extensions to the new science are important, like its wider ambitions to serve human life, but they should not obscure the fact that the most important social dimension of the new science was academic – concerned with the institutions which provided facilities and encouragement for this and other scholarly pursuits, and the

[7] Werner Sombart, *Luxury and capitalism* (English translation: Ann Arbor, 1967), esp. ch. 5; E. A. Wrigley, 'A simple model of London's importance in changing English society and economy 1650–1750', *Past and Present*, 37 (1967), 44–70.

[8] F. J. Fisher, 'The growth of London as a centre of conspicuous consumption in the sixteenth and seventeenth centuries', *T.R.H.S.*, 4th series 30 (1948), 37–50; Michael Foss, *The age of patronage: the arts in society 1660–1750* (London, 1971), ch. 4; Ellis,[149] ch. 4 and passim.

traditions and pressures that made up contemporary intellectual life. Indeed it is symptomatic of the crudity of much study of science's social relations that little attention has until recently been paid to such matters except through rather caricatured generalisation.

Misunderstanding has often been induced by the one novel feature of the Restoration, the advent of a specifically scientific institution, the Royal Society, which is often seen as synonymous with contemporary science and largely responsible for its achievement. Chapter 2 will consider the legitimacy of this, evaluating the Society and assessing its role in the organisation of intellectual life. Then chapter 6 will reconsider the position of the universities, for, whatever the importance of the new society, these were the chief centres of learning. The extent of their enthusiasm for and hostility to the new philosophy has been much debated, and this question will there be appraised in connection with the scholarly traditions of the time, whose vitality is sometimes underestimated. These can be (and were) argued to have had as significant 'social' functions as science, despite the disdain of some scientists and their latterday followers, and a vigorous controversy took place about the merits and implications of the new science which has long been familiar, if not always well understood.

There has been equal misapprehension about what contemporaries found perhaps most worrying in science's social relations, for a conviction of the close links between thought and action is manifest in the anxiety felt both within and without the scientific community about the danger of 'atheism'. The philosophical heterodoxy which had flourished in the Interregnum showed no sign of abating under the new regime and was widely considered subversive to moral values and social stability. Such fears had a marked effect on the public image and internal development of science, which will be surveyed in chapter 7: for while some saw scientific pursuits as conducive to infidelity, others argued that the study of God's handiwork was the best cure for unbelief.

The aim of this book is thus to use a close reading of manuscript and printed sources to show how Restoration science related to contemporary society in terms of support and apathy, facilities and impediments, motivation and reservations. To set the scene, however, we must first establish the character of science at the time. A proper appreciation of the broader role of Restoration science is impossible without understanding the nature and definition of science, the sometimes conflicting tendencies within it, and the forms in which it was most familiar to contemporaries – which are not necessarily the same as those in which it has been most familiar since. Only then can we understand developments during the Puritan Revolution, and, more important, the position which natural philosophy aspired to and achieved after 1660.

Restoration science: its character and origins

FRANCIS BACON AND THE SCIENTIFIC REVOLUTION

In 1668, Joseph Glanvill, Rector of Bath and a vigorous polemicist on behalf of the new science, published a book called *Plus Ultra*. The work was 'occasioned by a Conference with one of the *Notional* Way' – by which he meant the scholastic philosophy – that he had taken part in locally and to which he wished to give wider circulation by the printed word. The sub-title of the work proclaimed its theme: 'The Progress and Advancement of Knowledge Since the Days of *Aristotle*. In an Account of some of the most Remarkable Late Improvements of *Practical, Useful* Learning', and in it Glanvill described and defended the achievement of the science of his time. Thus it provides a helpful approach to the ideology of Restoration science, its definition and content, for Glanvill was in close contact with several leading researchers and was quite well-informed on the subject.

Perhaps most important is Glanvill's implicit definition of 'science', for it is essential to remember that this word was not then commonly employed in this connection and that its use invites misunderstanding: 'natural philosophy', or, to be more precise, 'the experimental philosophy', both have the more authentic ring of the seventeenth century. In fact it should be remembered throughout this book that 'science' and 'scientist' are used purely for convenience, to avoid constant recourse to more cumbersome words, and the reader should beware of anachronistic overtones due to the transformation of natural science in the centuries since.[1] The com-

---

[1] 'Science' is occasionally used in something approaching its modern meaning in the seventeenth century: see, e.g., Birch,[118] 1, 317. The commoner usage, however, is to mean 'Knowledge' and its branches generally: e.g., Glanvill,[66] sig. B6v, p. 25. See also below, p. 192.

moner contemporary phrase 'natural philosophy' is a clue to the correct status of the new science, as a department of philosophy from which it was only gradually separating itself. 'Experimental philosophy', on the other hand, best defines the type of intellectual activity that Glanvill set such store by, indicating how it was as much in method as subject-matter that the 'new' philosophers distinguished themselves from their predecessors.

Plus Ultra is also helpful as a reminder of how Glanvill and others conceived the aims of their enterprise. Glanvill called the learning that he was chronicling '*Practical*' and '*Useful*', and (like others) he was alive to its potential for increasing the '*conveniences of Life*' as well as 'for the *advancement* of *Knowledge*'.[2] Indeed this hope for the amelioration of life is intrinsic to the science of the time and almost all scientists considered technology germane to their interests, recognising the intellectual value of pursuits formerly considered rather menial. But they were aware of a distinction between pure and applied knowledge, and their concern for technology – which some considered more important than others – will be examined in a later chapter. The relative paucity of information on applied science in *Plus Ultra* illustrates how this, though not negligible, was peripheral to the intellectual revolution that Glanvill believed he was chronicling. He and others like him considered it their chief task to 'understand the *Artifice* of the *Omniscient Architect* in the composure of the *great* World, and *our selves*', and he dealt with a range of subjects very close to modern science – mathematics and astronomy, biology and anatomy, chemistry and statics, optics and pneumatics.[3] Theoretical investigations might be expected to have technological 'spin-offs', but these had to be kept subordinate to the more important job of reshaping ideas about the functioning of the natural world.

In all of the subjects he discussed, Glanvill and those whose achievements he described aspired to 'solid' knowledge: the most marked contrast they were conscious of was between the rapid and real advances in the natural sciences during the previous century, and the slow progress of knowledge that they associated with the former dominance of the philosophical synthesis evolved by the medieval schoolmen on the basis of the works of Aristotle. Whatever Aristotle's own scientific prowess, Glanvill attacked virulently the

[2] Glanvill,[66] p. 64.   [3] Glanvill,[66] p. 25 and passim. See below, ch. 4.

a priori 'notional' way of reasoning produced by the methods of teaching associated with scholasticism, which attached little significance to accurate observation and experiment or novel systematisation. Worse still was the common idolisation of Aristotle's achievement, which implicitly denied the possibility of any improvement of the synthesis that he had bequeathed.

Glanvill and his colleagues had a more optimistic view, and though there was an element of caricature in their simple contrast of the old and the new, this only enhanced its polemical strength. Nothing could be more striking than Glanvill's sense of the potentialities of the new science in its aspirations, its methods and its techniques. His aim (as he put it) was 'to encourage the *freer* and *better* disposed Spirits, to *vigour* and *endeavour* in the *pursuits* of *Knowledge*; and to raise the *capable* and *ingenuous*, from a *dull* and *drowsie acquiescence* in the *Discoveries* of former Times; by representing the great *Encouragements* we have to proceed, from *modern Helps* and *Advancements*'.[4] Most generally, he was referring to improvements of recent centuries like the printing press and the mariner's compass. More specifically, he referred to such novel techniques as the invention of logarithms and the improvement of scientific instruments which facilitated precise observation and mensuration – the microscope, the telescope, the thermometer, the barometer and the air-pump, all of which had been developed in the early and mid seventeenth century and were becoming increasingly common in Glanvill's time.

Glanvill went on to list some of the achievements of the new science over the previous century, both on the continent and in England, not least those which such novel techniques had made possible. He wrote in great detail about the astronomical discoveries of Johann Kepler and Galileo Galilei and their successors. He had much to say about the improvements of mathematical knowledge and their potentialities, and he drew attention to improvements in optics. He mentioned modern work in chemistry, 'by which *Nature* is *unwound*, and *resolv'd* into the *minute Rudiments* of its *Composition*', alluding to the work of Robert Boyle on this subject as also on statics and pneumatics.[5] He referred to advances in anatomy by continental and English researchers, laying stress on William Harvey's demonstration of the circulation of the blood and the

---

[4] Glanvill,[66] sig. B3. See also below, ch. 6. [5] Glanvill,[66] p. 11.

recent experiments on blood transfusion in London and Paris which were all the rage when Glanvill wrote his book. He also described recent achievements in 'natural history', in the collection of accurate information which he considered the essential basis for scientific advance: the most spectacular example of this was perhaps the detail about the structure of bodies revealed by the microscope in Robert Hooke's famous *Micrographia* (1665).

Glanvill's account is in some respects oddly partial, but it gives an idea of what scientific contemporaries saw as the major recent improvements of natural knowledge. It also conveys the excitement of discovery experienced by scientists and those on the fringes of scientific circles, their belief that 'There is an *inexhaustible variety* of *Treasure* which *Providence* hath lodged in Things, that to the Worlds end will afford *fresh Discoveries*', and their delight and amazement at the new phenomena that their investigations constantly brought to light.[6] Retrospectively, one could write a new *Plus Ultra* to take in not only the period up to 1668 but the remainder of the seventeenth century, and whereas in Glanvill's account the contribution of English researchers was impressive but proportionately modest, a consideration of the Restoration period as a whole would justify major claims for England as the most prolific contemporary centre for such studies and as the home of some of their greatest achievements. For the late seventeenth century was arguably one of the most fertile periods in the history of English science, amply fulfilling the promise felt by Glanvill. Its protagonists included a galaxy of well-known names, from Robert Boyle and Robert Hooke to Isaac Newton, John Ray and John Flamsteed, who made major advances on a whole range of scientific topics.

Boyle – whose achievement bulked larger than that of any other single figure in *Plus Ultra* – carried on with his experiments in statics and pneumatics and extended his attempts to vindicate a corpuscular theory of matter and to reformulate chemistry along mechanistic lines. Hooke – displaying an omnivorousness characteristic of the science of the age – worked out ideas on numerous subjects, many of them first adumbrated in *Micrographia*, from the nature of light and the theory of combustion to plant growth and gravitation. He also devoted much time to geology, and showed his exceptional fertility of mind in inventing mechanical devices throughout his career.

[6] Glanvill,[66] p. 7.

Hooke could also claim achievements in astronomical observation, and in this he was joined by such younger scientists as John Flamsteed and Edmond Halley, both of whom made major contributions to the recording and understanding of the movements of the heavenly bodies.

In medical science, the followers of William Harvey continued to advance the understanding of human physiology and embryology that he had pioneered, studying in detail numerous organs and functions in the body. Moving beyond Harvey's own concerns, followers like Thomas Willis, Richard Lower and John Mayow also attempted to understand the mechanics of respiration and the workings of the brain. Comparative anatomy attracted research from Edward Tyson and plant anatomy from Nehemiah Grew, while perhaps the greatest achievement of these years in the biological sciences was John Ray's, whose laborious classification of plants enabled him to draw up the most sophisticated system before that of Linnaeus. He also tried (with somewhat less success) to apply similar techniques to animals, birds and insects. This was an area in which other scientists were active, particularly Martin Lister, a pioneer in the classification and analysis of spiders.[7]

Undoubtedly the greatest achievement of the age, however, was not in the biological sciences but the physical. In his *Philosophiae Naturalis Principia Mathematica* (1687), Isaac Newton achieved the satisfactory geometrical account of the movements of the heavens to which scientists had aspired since the time of Galileo; he was also working in these years on research that was published in his rather different but no less crucial *Opticks* (1704). Newton built on the labours of others, but his transcendent importance was at once recognised. The *Principia* was widely acclaimed within a short time of its appearance, even by those unable to understand it, as 'the greatest highth of Knowledge that humane nature has yet arrived to'.[8] Indeed in providing a satisfactory astronomical system founded on the prevailing theories of the new science it had a profoundly important ideological role, while the *Opticks* had immense influence in the following century.[9]

[7] For a revaluation see Carr[142]; on Ray's work and its background see Raven[135]; on the other discoveries referred to in these paragraphs see below, pp. 199–201.

[8] Hunter,[143] p. 60. Cf. Axtell,[133] pp. 166–72.

[9] See below, ch. 7. For general accounts of eighteenth-century science see Arnold Thackray, *Atoms and powers* (Cambridge, Mass., 1970) and R. E. Schofield, *Mechanism and materialism* (Princeton, 1970).

Such a summary stresses the unity of the science of the age, and in doing so it echoes the similar sentiments of contemporaries. For when these sought to sum up all these diverse achievements they tended to underline, like Glanvill, the common basis of scientific endeavour in different fields in precise observation and experiment. As an early commentator on Newton's *Principia* put it, the 'experimental philosophy' of which that book was a classic example was distinguished by the fact that its protagonists 'assume nothing as a principle, that is not proved by phenomena . . . From some select phenomena they deduce by analysis the forces of Nature and the more simple laws of forces; and from thence by synthesis show the constitution of the rest.' This was echoed in the biological sciences by William Harvey's appeal to 'ocular inspection' of '*Nature* her selfe', and either view could be taken as a text for all these researches.[10]

In Restoration England there was a tendency to enunciate this principle by vociferously espousing its programmatic statement by Francis Bacon in a series of works published mainly in the 1620s. In his *New Organon* (whose title alludes to Aristotle's corpus of logical treatises, the *Organon*, which this was supposed to supersede) and the other component parts of his *magnum opus*, *The Great Instauration*, Bacon set out to provide a full-scale and systematic methodology for the reform of knowledge. He hoped to replace the sterile sophistries of scholasticism by a new programme of careful induction, involving the investigation of natural things by observation and experiment. In pursuit of this he advocated the use of inductive as against syllogistic logic to build 'axioms' concerning natural effects, and he urged the compilation of a 'natural history', a vast collection of material about natural phenomena to serve as 'the primary material of philosophy and the stuff and subject-matter of true induction'.[11] Bacon's Utopian sketch, the *New Atlantis*, published by his executors after his death, also proved deeply influential. This outlined his ideal of a publicly financed research institution, 'Solomon's House', for the co-operative study of God's works: its aim was 'knowledge of Causes, and secret motions of things; and the enlarging of the bounds of Human Empire, to the effecting of all things possible'.[12]

---

10 Cotes's preface in Newton,[63] pp. xx–xxi; William Harvey, *Anatomical Exercitations* (London, 1653), sigs. a5, a6v.
11 Bacon,[69] IV, 254.   12 Bacon,[69] III, 156.

By the late seventeenth century Bacon's name symbolised co-operative empirical philosophy and was on everyone's lips, 'celebrated as far as knowledge has any Empire'.[13] 'That inspired Heroe', was one evaluation; Glanvill wrote how Bacon had 'led the way to *substantial Wisdom*'; and a French contemporary had no doubt that he was 'the greatest Man for the Interest of Natural Philosophy that ever was'.[14] Such exalted praise was frequently echoed, and Bacon's importance to the early Royal Society was symbolised when he was enshrined as 'Artium Instaurator' in the frontispiece of the Society's first history.[15]

These assertions express a certain truth about the foundations of the new science, but they have often been interpreted too literally as an explanation of the seventeenth century's scientific achievement. Even if there was an element of unity in the experimental science of the time, it did not spring fully-fledged and without precedent from Bacon's writings, and recent scholarly debate about Bacon's role has resulted in a better understanding of the nature of the Scientific Revolution.

Even the devotion to careful observation and experiment that is often seen as most characteristically 'Baconian' had more varied roots. Empirical physiological research was pioneered in sixteenth-century Italy by Vesalius and others. Its most famous English champion was Harvey, who had a low view of Bacon and whose independent inductive research was carried out in a framework of modified Aristotelian ideas.[16] Another precursor was the sixteenth-century Swiss iatrochemist Paracelsus, who preceded Bacon not only in his challenge to the orthodoxy of Aristotle in philosophy and Galen in medicine but also in the empirical attention to chemical pathology that went with this.[17] Pre-Baconian empiricism can also be found in astronomy: Kepler's revisions to Copernicus's cosmological theories were based on his own observations and those of the Dane, Tycho Brahe, while an alert and anti-authoritarian astro-

[13] John Evelyn, trans. G. Naudé, *Instructions Concerning Erecting of a library* (London, 1661), sig. A5v.

[14] Fairfax to Oldenburg, 25 Jan. 1667, Oldenburg,[121] III, 316; Glanvill,[66] p. 75; Samuel Sorbière, *A Voyage to England* (English translation: London, 1709), p. 32.

[15] Sprat,[92] frontispiece. See cover illustration and appendix.

[16] Aubrey,[161] I, 299; Walter Pagel, *William Harvey's biological ideas* (Basel, 1967); Gweneth Whitteridge, *William Harvey and the circulation of the blood* (London, 1971), esp. ch. 1.

[17] Walter Pagel, *Paracelsus* (Basel, 1958); Debus,[43] chs. 1–2.

nomical tradition has also been discerned in sixteenth-century England.[18] At best Bacon only gave a systematic statement of an approach that already existed.

More important, perhaps, is the question of whether inductivism was the most important agent of early modern scientific change. In the history of science as a whole, few are now satisfied with the idea that knowledge has advanced by the simple accumulation of correct data, as if this led almost inevitably to improved theories. Instead, it may be argued that of itself precise observation produced only a heap of miscellaneous information: more significant was the body of theory with which such data was approached and the way in which this was modified by new hypotheses. For these, though often at first put forward without any experimental support, could be refined and tested to become the basis of new explanations of how the world worked. According to such views Bacon was 'completely negligible' in the Scientific Revolution.[19] What mattered more was not a general programme like his but novel theories and explanatory possibilities suggested by men like Galileo and Descartes, for these drew attention to new problems for research and hence founded quite precise 'research traditions' which set the terms of reference in different subjects.

In the biological sciences, one such tradition was the physiological investigation of which Harvey and his followers were exponents; another was the taxonomy to which John Ray contributed, with its roots in Renaissance Europe. In chemistry the views of Paracelsus offered an explanatory system based on the reactions of three principles, which was modified and extended by seventeenth-century thinkers like J. B. van Helmont. More important was the reshaping of physics and astronomy associated especially with Galileo, which has often been seen as the heart of the Scientific Revolution. Galileo propagated a pervasive curiosity about physical and astronomical problems, while the challenge of finding a geometrical resolution of the nature of celestial and terrestrial motion preoccupied scientists both on the continent and in England. In these and other topics — from statics and optics to geology — English scientists reacted to problems tackled and hypotheses promulgated by writers at home and abroad, and it is in these terms that their achievement is to be

[18] Alexandre Koyré, *The astronomical revolution* (English translation: London, 1973), pt. 2, sect. 2; F. R. Johnson, *Astronomical thought in Renaissance England* (Baltimore, 1937). [19] Koyré,[25] p. 1.

understood. This is shown clearly by the recent studies cited in the bibliographical essay at the end of this book.

What is more, in the physical sciences mathematics was integral to the formulation of new theories, and it is possible (as some have shown) to write a history of the Scientific Revolution in primarily mathematical terms: this again indicates the limitations of too exclusively a Baconian view of it, since Bacon was not very mathematically minded. Many saw mathematics as the true centre of the new philosophy, building their methodology around it, and some of the most critical scientific innovations of the period were both conceived and worked out mathematically. The belief that everything could be reduced to mathematics is characteristic of the thought of the time and it was taken to absurd extremes, as when efforts were made to reduce politics and even morality to geometry: thus the Dublin virtuoso, Samuel Foley, wrote a paper entitled 'Computatio Universalis seu Logica Rerum, Being an Essay attempting in the Geometrical Method to Demonstrate a Universal Standard whereby one may judge the real Value of Everything in the World'.[20]

Mathematics was central to perhaps the most influential doctrine of the Scientific Revolution, the so-called 'mechanical philosophy', the view that all natural effects could be reduced to the simple interaction of units of matter in motion. This contrasted both with the views of Aristotle, who had postulated a more complicated relationship of elements and contraries, and also with new cosmologies of the sixteenth century like that of Paracelsus, and it was another innovation to which Bacon was a stranger. Two men were chiefly responsible for the new viewpoint, the Frenchman Descartes and the Englishman Thomas Hobbes, both of whom devised on this basis complete and exhilarating philosophical systems. Indeed Hobbes so rigorously drew from this philosophical position all of its logical corollaries that he made it quite embarrassing for scientists in ways that will become apparent in chapter 7. Descartes, however, was more significant than Hobbes in applying these ideas to a wide range of scientific problems and defining crucial areas of research in which others followed. Much of the intellectual effort of Restoration science was devoted to testing and refining Cartesian mechanical theories in new fields of research, from biology to chemistry.

But if it is misleading to accept Baconian platitudes as a complete

[20] Bredvold,[39] esp. pp. 165–72; Hoppen,[117] pp. 122–3.

account of seventeenth-century science, it is equally wrong to imply that these more varied sources of innovation excluded Baconianism. For though Baconian induction has sometimes been seen as completely antipathetic to hypotheses, this misrepresents the position. Bacon's philosophy provided a general programme into which a whole range of particular ideas from other sources could fit, hypotheses being quite acceptable so long as they were seen as such and only espoused after rigorous testing. Such sophisticated Baconians as Boyle and Hooke in the Restoration realised that what was needed was a judicious balance between a generalised induction and particular hypotheses, and this was echoed in lesser men too, who cited Bacon's own example to allow themselves to digress from description 'ever and anon into the *Ætiology*' of phenomena.[21]

Baconianism did, however, create some tension. Late seventeenth-century scientists, particularly in England, were profoundly Baconian in their insistence on the need for proof of novel theories rather than mere plausibility. They rightly saw that some new views of the workings of the world, including Descartes's, were in danger of being as *a priori* as the old scholastic theories which they attacked, and Boyle pilloried the 'thought' experiments that abstract theoreticians often devised to back up their ideas, asking (of a suggestion by Blaise Pascal) how an experimenter was supposed to stay twenty feet under water to make the precise readings that he postulated.[22] This led to some strain in English scientific circles, for some were more Baconian than others. Thus the Halifax doctor and naturalist, Henry Power, displayed an enthusiasm for Cartesian systematisation that Boyle regarded as premature, seeing the real danger that excessive deductive rationalism might prove as stifling a dogma as the old scholasticism which the new science was superseding.[23]

It is also clear that Baconianism encouraged indiscriminate collecting of information relevant to no particular hypothesis. This was inevitable in some subjects, such as the study of heat, where there was no research tradition to draw on; it was encouraged by the new exactitude of observation made possible by scientific instruments which justified experiment almost for its own sake. But Baconianism also encouraged random collecting in subjects where it was unconstructive, enabling those lacking in imagination to make a virtue of

[21] Joshua Childrey, *Britannia Baconica* (London, 1661), sig. B2.
[22] Boyle,[71] II, 745–6.   [23] Webster,[29] pp. 167–8.

their abstention from philosophical speculation.[24] The charter for such activity was provided by Bacon's idea of a 'natural history', the store of data from which hypotheses were to be inductively derived, perhaps his most significant methodological legacy to seventeenth-century science. Even Hooke stated that 'until this Repository be pretty well stored with choice and sound Materials, the Work of raising new Axiomes or Theories is not to be attempted', and lesser men took Bacon at his word. They proceeded to write natural histories of an almost entirely miscellaneous kind, taking their inspiration from Bacon's own contribution to the genre, his *Sylva Sylvarum* – itself very much such a compilation – rather than from the *New Organon*. Perhaps the best example of this is the *Natural History of Oxford-shire* (1677) of Robert Plot, later secretary of the Royal Society and first curator of the Ashmolean Museum at Oxford, a work that was both popular and demonstrably influential.[25] Indeed many have overlooked the implication of this interpretation of Bacon's philosophy for the contemporary image of science and for science's social extension, which will be examined in chapter 3.

In its distrust of dogmatism and its stress on the importance of open-mindedness even on seemingly obvious questions Baconianism had a further effect on the science of the time, for it encouraged a tolerance of theoretical diversity and this explains another feature of Restoration science which many have found puzzling. In England, as elsewhere, it is easy to find vociferous defences of mechanistic explanations like Boyle's in his 'Of the Excellency and Grounds of the Corpuscular or Mechanical Philosophy'.[26] But these had to co-exist with a wide range of potentially contradictory philosophies, often with a strong occult tinge, philosophies which even mechanists took seriously with a Baconian insistence on the need to test their theories and conclusions.[27] Boyle was only too aware that his experiments would 'Occasion among the *Virtuosi* several *Quere's* and *Conjectures* according to the differing *Hypotheses* and *Inquisitions*, to which men are inclined', while Martin Lister urged that everybody should be 'left to his own sense and thoughts concerning the manner of understanding the causes of things; which

[24] Kuhn,[26] pp. 46–7; Carr,[142] p. 380.
[25] Hooke,[65] p. 18; Hunter,[143] ch. 2.   [26] Boyle,[71] IV, 68–78.
[27] See, for instance, Charles Webster, 'Water as the ultimate principle of nature: the background to Boyle's Sceptical Chymist', *Ambix*, 13 (1966), 96–107.

I think, however different, will never injure philosophy or real friendship'.[28]

The 'chemical philosophy' of Paracelsus retained its vitality in the late seventeenth century, and it is also possible to trace the residual influence of the 'Hermetic' tradition so popular in the sixteenth century, when it was believed that the writings of Hermes Trismegistus were of the utmost antiquity, though it is now known that they are of early Christian date. Both Paracelsian and Hermetic writings attributed significance to occult forces, to sympathy and antipathy, and Restoration scientific circles were filled with men who combined vocal devotion to Baconian empiricism with confused world-views full of astral influences and non-mechanical causes and cures.[29] This was encouraged by the empirical element which these traditions shared with Baconianism, while attention to Bacon's writings could itself leave traces of his 'semi-Paracelsian' world-view.[30] Vitalist theories also received a fillip from the Platonic and neo-Platonic ideas popularised in the mid seventeenth century by the Cambridge Platonists and others.

Even occult pursuits like astrology and alchemy were espoused, sometimes publicly, as in the case of Plot, who hoped to gain endowment for a college dedicated to the discovery of the elixir of life.[31] This shows that the devotees of these activities saw nothing inconsistent about their commitment both to them and to the new science, while others wished to systematise astrology along Baconian lines so that it could aspire to the status of the new science. Not surprisingly, therefore, those who attacked the new philosophy tended to conflate scientific and occult pursuits, though this has sometimes perplexed historians.[32]

It might be thought that this represented a legacy from the past among lesser men, on which more original scientists indulgently smiled. Certainly Hooke regarded occult-tinged explanations of natural phenomena as 'Strange whimsys' and sought to replace them

---

[28] *Phil. Trans.*, 2 (1667), 592; Lister to Aston, 14 Dec. 1683, Birch,[118] IV, 244 (misdated). [29] Debus,[43] chs. 5, 7; Hunter,[143] ch. 2.

[30] Graham Rees, 'Francis Bacon's semi-Paracelsian cosmology and the *Great Instauration*', *Ambix*, 22 (1975), 81–101, 161–73; 'The fate of Bacon's cosmology in the seventeenth century', *Ambix*, 24 (1977), 27–38.

[31] Gunther,[129] pp. 411–13; F. Sherwood Taylor, 'Alchemical papers of Dr Robert Plot', *Ambix*, 4 (1949), 67–76 (though his suggestion that Plot felt he ought to keep his alchemy secret (p. 76) is implausible: cf. Hoppen,[55] p. 261).

[32] Hunter,[143] pp. 117–21, 136–7, 144–5; cf. Capp,[152] pp. 183–6.

with '*Motion, Figure,* and *Magnitude*'.[33] But similar views can be found in lesser men who wrote off sympathy and antipathy and occult qualities as 'nothing but a refuge of ignorance', while major scientists concerned themselves with the occult.[34] The most spectacular case of this is Newton, whose alchemical experiments (though the subject of much controversy) undoubtedly went beyond any mechanistic rationale. He even warned Boyle not to meddle in matters which 'may possibly be an inlet to something more noble, not to be communicated without immense dammage to the world if there should be any verity in the Hermetick writers', and it has been claimed that the *Principia* represented only a brief interlude in his working career whereas alchemy was a longer-lived interest.[35] Though his achievement in the *Principia* depended on a clear mathematical demonstration of the system of the world, the central idea of force which it invoked to complete its picture may well have had its roots in the occult tradition and was attacked by more thoroughgoing mechanists.[36]

So the explanatory systems of Restoration scientists varied, as did their method. But it is not paradoxical to claim that Restoration science was characterised partly by its very tolerance of diversity, which was itself a unifying factor. It is true that during the period some attempts were made to assert an orthodoxy both in method and theory: but their success was limited.[37] In general it is almost impossible to categorise science into different varieties since contrasts are easily overrated, especially by those in search of facile external correlations. Thus Newton, Ray and Boyle had much in common as thinkers, sharing many similar views and preoccupations. But it would be hard to find a contrast greater than that between the science of Newton and Ray, the former concerned with geometrical abstraction and the most complete of cosmological theories, the latter with the collection of endless precise botanical information. Whereas for Newton and others mathematics was essential to the new science, Ray and his like had scanty knowledge of mathematics and little place for it. Ray also stood apart from Boyle in the physical sciences in his lack of interest in performing

---

[33] Hooke,[165] pp. 331, 335, 337–9 (concerning Oliver Hill: cf. Birch,[118] III, 363, 366–7, 371); Hooke,[64] sig. g1.

[34] 'The Artifice of Alum', Royal Society MS Extra I, item 2, fol. [4]. On the author see below, p. 108 n. 84.

[35] Newton to Oldenburg, 26 April 1676, Newton,[127] II, 2; Westfall,[59] 195–6.

[36] See below, p. 185. [37] See below, ch. 7.

experiments, but Boyle stood with Ray in contrast to Newton in his relative ignorance of complex mathematics.

Such contrasts between different fields are predictable, and they only reinforce the essential unity of the science of the time. Indeed the best of recent scholarship has restored to prominence the common character of scientific research obscured by too much disembodied stress on methodology and too much sensationalist attention to occultism. This is the sense of intellectual discovery in natural philosophy, the formulation of hypotheses and their testing against experimental data in all sorts of different but overlapping subjects.[38] Scientists in the late seventeenth century worked together and felt committed to a common task of reforming and extending knowledge, even if they differed about the best means to achieve this. They shared a sense of achievement like that expressed by Glanvill in *Plus Ultra*.

THE LEGACY OF THE INTERREGNUM

The formative years for the scientific tradition characterised in the last few pages were those of the Puritan Revolution and the Interregnum. Scientific activity had, of course, gone on before. Bacon and Harvey have already been mentioned, as have the Elizabethan astronomical writers, who found heirs in the early seventeenth century; there was also some interest in Paracelsian ideas.[39] But, though research will certainly bring to light more pioneers earlier in the century, there was undoubtedly an explosion of science in its middle decades. When John Aubrey looked back after the Restoration, he thought that 'the searching into Naturall knowledge began but since or about the death of King Charles the first' in 1649, and there is truth as well as oversimplification in this.[40] Baconianism now first became a commonplace slogan, thanks perhaps above all to the Prussian emigré Samuel Hartlib and his circle, whose indefatigable promotion of Baconian and other schemes is surveyed in Charles Webster's monumental book, *The Great Instauration* (1975). At a rather different level, the 1640s saw the emergence of a school of 'Harveians', who took up ideas that Harvey had pioneered and extended them, part of a renaissance of medical teaching and

[38] For a good specimen of such work see Frank.[40]
[39] A. G. Debus, *The English Paracelsians* (London, 1965).
[40] Aubrey 3, fol. 18v.

research at Oxford. There was also a great increase of interest in
Paracelsian ideas, beside which earlier curiosity seems minor: a
number of translations were published and Paracelsian doctrine now
first reached wide circulation. At the same time, the ideas of Galileo,
Descartes and other new philosophers were introduced to a larger
English audience through the efforts of popularisers like John
Wilkins. From about 1650 even the yearly almanacs that had been
purveyed throughout the early modern period show much more
awareness than hitherto of the new astronomy of Copernicus,
Kepler and Galileo.[41]

Moreover many English scientists who published important
findings after 1660 – Boyle, Hooke, Ray, Willis and others,
whose importance has been indicated – came to maturity and
laid the foundations of their later work in the Interregnum. Their
investigations formed part of a great expansion in scientific
work, of experimental inquiries into topics like astronomy
and optics along lines suggested by Galileo and Descartes, and of
studies of more miscellaneous subjects from anatomy and biology
to chemistry.

It was encouraged by the emergence of active centres of research
and discussion. Perhaps the most famous of these was the group
which met in London in 1645 and took an interest in a whole range
of problems, largely from continental science: this is shown by a
subsequent report by John Wallis, one of its active members and
later Savilian Professor of Geometry at Oxford. Wallis's account
gives a sense of Englishmen familiarising themselves with new and
exciting matters in which they were later to emulate their European
exemplars, and there is also evidence for a cross-fertilisation in the
group between mathematicians and medical men which was full of
promise for the future.[42] These meetings were affected by the exodus
of many of the group's members to Oxford at the end of the 1640s,
but they evidently continued throughout the Interregnum. In the
1650s the London College of Physicians was also a flourishing centre
of research: one contemporary likened it to 'Solomon's House' in
Bacon's *New Atlantis*, itemising its Fellows' achievements in subjects from comparative anatomy to metallurgy and even brewing.
In addition, activities went on in smaller and less official groups

[41] Frank,[40] esp. ch. 2; Webster,[72] pp. 273–82; Nicolson,[153] pp. 8, 11–12, 18–19.
[42] C. J. Scriba, 'The autobiography of John Wallis, F.R.S.', *N.&R.R.S.*, 25 (1970),
39–40; Webster,[72] pp. 54–7; Frank,[40] ch. 2.

elsewhere in the metropolis, not least among Hartlib's associates.[43]

This was paralleled at Oxford. The medical research stimulated by Harvey flourished throughout the Interregnum, and natural philosophy was also pursued in small college groups. But scientific activity was intensified in the 1650s by a more formal group which even aspired to institutional status and which seems to have originated when numerous intellectuals from London colonised Oxford at the end of the 1640s. In the 1650s it was associated most of all with John Wilkins, Warden of Wadham College and a great enthusiast for natural philosophy. Those active in Oxford at this time included John Wallis, Seth Ward, Christopher Wren, William Petty, Thomas Willis and (in the later years of the Interregnum) Robert Boyle. They pursued scientific investigations which did much to give English science an international reputation, preparing the way for the achievements of the Restoration in mathematics, statics, optics, astronomy and chemistry. They also pioneered the application of instrumentation to science, experimenting with the use of telescopes and microscopes, while experiments were carried out in blood transfusion.[44] Oxford science looked forward to the Restoration not least in grafting Baconianism together with the sophisticated mathematical traditions of continental science in the manner typical of the later years of the century. It was also now that Boyle was at work trying to provide an experimental basis for the corpuscularian ideas that he derived from French thinkers.[45]

There was a more muted analogy to this at Cambridge. Among scientific authors of the Restoration, those who were at Cambridge in the Interregnum and developed their scientific interests there included Henry Power and John Ray. Activities included dissections, chemical experiments and simpling expeditions, while there was also a good deal of interest in the philosophy of Descartes, especially under the auspices of Henry More. In the 1650s More expounded a blend of Cartesian ideas with the characteristic vitalism of the 'Cambridge Platonism' of which he was a prominent exponent and this was another formative influence on Restoration science, not least

[43] Webster,[72] pp. 57, 93 and passim; Webster,[73] pp. 398–407.
[44] In addition to Frank,[40, 75] see Webster,[72] pp. 154–72, and Bennett,[141] chs. 2, 3 and 5. On Wilkins there is an excellent account by Hans Aarsleff in Gillespie,[134] XIV, 361–81.
[45] Wood,[28] pp. 214–22; Hesse,[33] esp. p. 79; Kargon,[50] ch. 9.

on Newton.[46] Indeed a contrast between Cambridge and Oxford
alumni suggests a sociological explanation of at least some of the differences in scientists' interests and approaches after 1660. Cambridge produced more men interested in non-human biology than
Oxford; Power's more overt Cartesianism may also have stemmed
from his Cambridge background, in contrast to the greater emphasis
on Baconianism at Oxford; while contrasting views on vitalism will
be surveyed in chapter 7.[47]

How significant is it that science flourished in this way in the years of
revolutionary upheaval? Though it is one of the most controversial
questions in the history of seventeenth-century science, this vexed
matter must be clarified as an essential background to developments
in new circumstances after 1660. Of attempts to posit a direct
connection between science and the Puritan Revolution the most
recent and learned has been Charles Webster's in *The Great Instauration*. The kernel of his book – and perhaps the best focus for
such efforts – is the Hartlib circle, which was closely allied to the
parliamentary cause in the 1640s and to the republican regime
thereafter. Hartlib's associates included many strongly committed
both to the advancement of learning and to reform in numerous
areas of national life; they combined such aspirations into a millen-
arian scenario echoing that which underpinned the Puritan Revolution as a whole.[48]

Hartlib patronised and encouraged various young intellectuals
including Boyle and William Petty, and, whatever his own scientific
prowess, his enthusiasm was stimulating and productive. Furthermore he undoubtedly contributed to the scientific vitality of the
day by his endless curiosity and his sedulity as a correspondent
linking together numerous widely-scattered intellectuals. Since the
Long Parliament patronised Hartlib and the parliamentary regime
paid him a pension, the Puritans in power may take part of the credit
for his activity, while some of the projects which Hartlib promoted
paid dividends to the revolutionary regime. In the late 1640s his
friends were engaged in a saltpetre project to increase the supply of

---

[46] Webster,[72] pp. 144–53; Webster,[29] pp. 152–6. For a caveat see Carr,[142] p. 11. See
also below, pp. 181–3.   [47] Wood,[28] ch. 4.
[48] Webster,[72] pp. 1–31. Cf. K. R. Firth, *The apocalyptic tradition in Reformation
Britain* (Oxford, 1979), ch. 7. See also William Lamont, *Godly rule* (London, 1969)
and Paul Christianson, *Reformers and Babylon* (Toronto, 1978).

gunpowder for the parliamentary armies, while in the 1650s Petty produced his great 'Down Survey' of Ireland, which was an essential adjunct of the Cromwellian settlement there. Since this grew directly out of the Baconian programme of the Hartlib circle, it illustrates a connection between their concerns and the preoccupations of the regime, as do Hartlib's (largely unrealised) plans for reform in medicine and education, which combined proposals for modernisation with an attack on restrictive monopolies.[49]

The parliamentarians also contributed to the scientific vitality of the universities by intruding their candidates into academic posts at Oxford and Cambridge: Wilkins was himself a parliamentary nominee. A link has also been asserted between Puritan radicalism and the new popularity of the occult cosmologies of the sixteenth century. Those who now published Paracelsus's works in English for the first time included several with clear radical associations, and many scented a connection between mystical science and spiritual reform.[50]

Apart from these direct links between science and the ideology and policy of the Puritan Revolution, however, science also gained from new circumstances which the Revolution inaugurated but which were largely accidental. Thus the effective freeing of the presses by the war and its aftermath – which the parliamentary regime did not welcome but could do nothing about – clearly played its part in arousing intellectual curiosity and enhancing scientific receptiveness. Conservatives complained about the 'intollerable scab of translations', while 'for Judiciary Astrologers, Conjurers, Lyrick moderne poets, Dispensatories, etc. all this is levelling, and laying all common'.[51]

Indeed many of the beneficial effects of the revolutionary upheaval on science were fortuitous. Whereas some intellectuals had their careers advanced by the parliamentary regime, science gained because others had theirs thwarted. More than one Anglican was deflected by the Puritan triumph from a clerical to a scientific career, while for others there ensued a period of enforced leisure favourable to scientific pursuits, as in the case of the mathematician Edmund

---

[49] Webster,[72] pp. 61, 71, 428–44; Sharp,[144] pp. 116–36; T. C. Barnard, *Cromwellian Ireland* (London, 1975), ch. 8; Webster,[72] chs. 3, 4.
[50] Webster,[72] pp. 279–81; Rattansi,[80] pp. 25–8, 29; Hill,[78] ch. 14.
[51] Jaspar Needham to Evelyn, 5 April 1649, Christ Church, Oxford, Evelyn MS 3.2, fol. 93.

Wingate.[52] More important, the effect of the war was deeply unsett-
ling for all, and the intellectual ferment in its aftermath affected not
only parliamentarians – as is familiar from the work of Christopher
Hill and others – but also their royalist opponents, who sought new
sources of opinion and authority. Thus the Anglican, Thomas
Vaughan, turned to Hermeticism in despair after the execution of
Charles I in 1649 and his eviction from his parish; other royalists
espoused the novel doctrines of Hobbes, while others took up
Baconianism.[53]

It is therefore not surprising that many of the most active scientists
of the Interregnum were neither Puritans nor collaborators with the
regime. Harvey's group flourished in Oxford when it was controlled
by the royalists in the 1640s, and among his most notable followers
was Thomas Willis, who in the 1650s combined important research
with crypto-Anglican activities.[54] Other royalist scientists included
Seth Ward and Christopher Wren, both of them associated with the
physician Sir Charles Scarburgh, whose house was 'the Rendesvous
of most of the learned Men about *London*, especially of those of the
Royal Party' in the late 1640s.[55] Hartlib also had royalist supporters,
including Sir Justinian Isham, Sir Christopher Hatton and John
Evelyn, while royalist enthusiasts for the Hermetic philosophy included Elias Ashmole as well as Vaughan. The most influential
mathematician of these years, William Oughtred, was so strong a
royalist that he died of a heart attack at his delight on hearing of the
Restoration.[56]

There were some contrasts between scientists who collaborated
with the regime and those who did not. It has been argued that the
champions of the Harveian tradition tended to be Anglicans and
royalists while Hartlib and others associated with the parliamentary
regime devoted proportionately more time than their conservative

52 Webster,[72] p. 142; Taylor,[156] p. 205.
53 F. B. Burnham, 'The More-Vaughan controversy: the revolt against philosophical
enthusiasm', *J.H.I.*, 35 (1974), 39; Samuel Tuke to Evelyn, 25 Oct. 1650, Evelyn
Corr., no. 1277; Baconianism will be dealt with in my forthcoming study of John
Evelyn.
54 Webster,[72] p. 130; Frank,[40] ch. 2; Anthony Wood, *The history and antiquities of the
University of Oxford*, II (Oxford, 1796), 613.
55 Walter Pope, *The life of Seth, Lord Bishop of Salisbury*, ed. J. B. Bamborough
(Oxford, 1961), p. 124.
56 H. R. Trevor-Roper, 'Three foreigners: the philosophers of the Puritan Revolution'
in *Religion, the Reformation and social change* (London, 1967), 256; Webster,[72]
p. 282; Aubrey,[161] II, 111–12.

colleagues to the technological matters that will be surveyed in chapter 4.[57] But such attempts to split up the science of the time easily end in caricature, obscuring how much its devotees had in common. Scientists in Oxford concerned themselves with technology as well as almost the whole spectrum of science, though some authors have seen this group as 'academic' in contrast with their practical London contemporaries. Similarly the College of Physicians was a centre not only of Harveian physiology but of other scientific pursuits, including some – like metallurgy and chemistry – overlapping with the characteristic concerns of Hartlib and his friends.[58] Though emphases naturally differed, science already had a unity which could overcome party loyalties, and so it has been suggested that the group that met in London in the 1640s was self-consciously 'neutral' because of the strength of members' feelings on non-scientific topics.[59]

Indeed from the late 1640s a positive virtue was made of the way in which science transcended parties. The context of this was the growing revulsion of moderate opinion against the extremes of political and religious radicalism in these years, which went far beyond the assumptions and hopes of the idealists who had opposed the King before the Civil War. Especially alarming was the proliferation of warring sects; this was widely blamed on religious dogmatism and mysticism, and many reacted by seeking a fresh basis for faith in consensus rather than illuminism, in rationalism rather than dogma. They thus took up a revolt against narrow credalism that had already had a limited circulation on the continent and in England, notably in the so-called 'Tew Circle', and which the experience of sectarianism made all the more attractive.[60] 'It is strange, that men should rather be quarrelling for a few trifling opinions, wherein they dissent, than to embrace one another for those many fundamental truths, wherein they agree', Boyle felt, and he and others sought to lay the foundations of a religious compromise, in which they were convinced that science could play a part through the humble pursuit of knowledge of God's works.[61]

Even the Hartlib circle fits into this context, despite its commitment to reform. Hartlib's outlook was essentially moderate, and he

[57] Webster,[72] p. 498.
[58] Frank,[75] p. 208; Webster,[73] pp. 398–407.   [59] Webster,[72] p. 55.
[60] McAdoo,[87] ch. 1; McLachlan,[294] chs. 4–5; Colie,[296] ch. 1.
[61] Boyle to Dury, 3 May 1647, Boyle,[71] I, xxxix.

and his associate, the itinerant evangelist John Dury, were obsessed by the search for religious unity based on the fundamentals of a reasonable faith. When practical attempts at church unity failed, the Hartlibians laid increasing stress on practical piety and the pursuit of natural knowledge as the means to true enlightenment.[62] A reaction against dogma and sectarianism also underlies the theology of the Cambridge Platonists: this too had originated earlier but was brought into sharper focus by the experience of the 1640s. Henry More attacked those who 'have made their own inventions and argumentative conclusions articles of faith', while his friend Ralph Cudworth's purpose 'was not to contend for *this* or *that Opinion*; but onely to perswade men to the *Life of Christ*, as the Pith and Kernel of all Religion'.[63]

The Oxford scientists show a similar pattern, for they too were moderate and hostile to radical extremes. Though John Wilkins had been intruded into Wadham, his open-mindedness was proverbial. Many country gentlemen, especially of the royalist party, sent their sons to be educated under him, while Robert Hooke considered Wilkins 'an evident Instance, what the true and the *primitive unpassionate Religion* was, before it was *sowred* by particular *Factions*'.[64] Wilkins and his Oxford associates espoused an undogmatic theology, which stressed the need for consensus, piety and good works: in other words, the 'Latitudinarianism' that he and others were to champion after the Restoration. Moreover this was combined with the conviction that the study of the natural world made manifest God's glory, that scientific investigation could cool religious controversy, ending empty verbal disputes and manifesting the goodness of God in his creation. Traditional arguments of natural theology now received a new force and cogency from their exposition in scientific circles.[65]

The reaction against sectarianism and appeal to moderation had some effect on scientific theory. Henry More was vociferous against 'enthusiasm', and his zealous espousal of Cartesian ideas owed something to this. Similarly, the royalist doctor, Walter Charleton,

---

[62] Batten,[77] pp. 131–3 and passim; John Dury, *An Earnest Plea for Gospel-Communion In the Way of Godliness* (London, 1654).

[63] McAdoo,[87] p. 118; Ralph Cudworth, *A Sermon Preached before the Honourable House of Commons* (Cambridge, 1647), sig. ¶3.

[64] Pope, *The life of Seth*, p. 29 (there is no evidence that this occurred only late in the Interregnum, as claimed in Webster,[72] p. 85); Hooke,[64] sig. g2.

[65] Shapiro,[85] ch. 3 and passim; McAdoo,[87] ch. 6.

who had eagerly taken up the ideas of Paracelsus in the 1640s, reacted against this – evidently alarmed by their radical associations – and instead popularised continental atomist ideas.[66] Boyle was also critical of the use made of Paracelsian ideas by the sectaries, and it has been argued that a corpuscularian orthodoxy emerged in the 1650s as a reaction to this threat, though the theoretical diversity surveyed in the first part of this chapter should indicate the danger of over-simplification in this.[67]

Perhaps more important, the appeal to moderation and consensus strengthened the hold of Baconianism in scientific circles. Whatever the actual role of theory in science, there were strong incentives to stress the importance of laborious, co-operative induction as the proper way forward. Boyle drew up some of his most explicit statements of inductive method at this time, arguing 'that numerous Observations of Sense ought to be diligently sought after and procured' before conclusions were drawn.[68] In these circumstances, the Baconian flavour of English science was enhanced as part of a wider ideology which stressed the advantages of science in reconciling warring factions by appealing to all reasonable men, and, at the same time, in offering improved knowledge of the natural world from which technological aids might derive, in contrast to the fruitlessness of earlier intellectual traditions.

The most complete and cogent exposition of this viewpoint was written after the Restoration, but it grows directly out of the experience of the Interregnum and it echoes statements along similar lines during the preceding decade. This is Thomas Sprat's *History of the Royal Society*, published in 1667 though begun in 1663, one of the most often quoted and frequently misunderstood works of its time. Sprat's book is as much a confession of faith as a factual record, and reservations have rightly been expressed about the extent to which it is trustworthy as a historical source or as a statement of the 'official' policy of the body it celebrated. What is certain, however, is its significance as a summation of polemical attempts to define the role of science in the atmosphere of the mid seventeenth century. Sprat had been an undergraduate at Wadham under Wilkins, and Wilkins was closely associated with the book while it was being prepared; it was also vetted by prominent Fellows of the Society,

---

[66] Rattansi,[80] pp. 26, 28, 30–1; N. R. Gelbart, 'The intellectual development of Walter Charleton', *Ambix*, 18 (1971), 149–68.

[67] Rattansi,[81] pp. 19–23; Jacob,[83] pp. 110–11.   [68] Westfall,[30] p. 66.

who clearly agreed with the work's main thrust even if they cannot be held responsible for its details.[69]

In his philosophy, Sprat preached a familiar, empirical, anti-dogmatic viewpoint: indeed, in a deliberate attempt to commend the new philosophy to a wide public, he went further than most in stressing the need to collect information and playing down the role of hypotheses.[70] Sprat attached particular value to 'the *Relations* of the effects of *Nature*, and *Art*' which the early Royal Society received in 'infinite' number from men of diverse ability, several of which were printed in the *History*. He also made much of the role as arbiters of scientific truth of 'plain, diligent, and laborious observers: such, who, though they bring not much knowledg, yet bring their hands, and their eyes uncorrupted'.[71] He thus typically expressed the distrust of the speculative intellect embodied in Baconianism, while collaborative research was also valued for counteracting the fragmentation of English society which dogmatism and sectarianism were believed to have encouraged.

Sprat asserted the antipathy between natural philosophy and 'the passions, and madness of that dismal Age', the Interregnum, feeling bound to apologise even for the word 'committee' because of the distasteful overtones it had acquired from the use of committees as agencies of parliamentary authority.[72] He similarly attacked illuminist religion, arguing instead for a rational and straightforward creed that would be acceptable to all. In science he attacked both 'Modern Dogmatists' and enthusiasts: he was especially outspoken against alchemists, whose 'Writers involve them in such darkness; that I scarce know, which was the greater task, to understand their meaning, or to effect it'.[73] The two parts of Sprat's case came together when he argued that since 'a *rational Religion*' was the 'universal Disposition' of the age, the philosophical attitudes and activities of the new science should properly be at the heart of the Anglican church. He also emphasised that the scientist 'has always before his eys the *beauty, contrivance*, and *order* of *Gods Works*'.[74]

The moderation that Sprat and his colleagues proclaimed did not (as has sometimes been thought) make them withdraw from wordly matters, particularly if it seemed possible that a consensus might be

[69] Wood;[93] Aarsleff in Gillespie,[134] XII, 580, 584 and 580–7, passim.

[70] Wood,[93] esp. p. 12.

[71] Sprat,[92] pp. 195, 72. [72] Sprat,[92] pp. 53, 85.

[73] Sprat,[92] pp. 28, 37. [74] Sprat,[92] pp. 374, 349.

achieved in mundane as well as intellectual affairs. Arguing that
science was at one with 'the present prevailing *Genius* of the *English*
Nation', Sprat enthusiastically listed the classes and professions
from whom he hoped Baconian science would derive support and to
whom (he argued) it offered no threat.[75] Churchmen and scholars,
merchants and artisans, statesmen and diplomats, and above all the
gentlemen who formed the backbone of the political nation: all
could work together, each contributing their share and each deriving
benefit. Sprat and others with similar views had visions of a Golden
Age of political stability, economic prosperity and intellectual
advancement which would satisfy all rational men. His book was
filled with fervour for national improvement through trade, colonial-
ism and industry, and this was the scenario he sketched for the
triumphant progress of science. Indeed in his hopes for the close
connection of science, commerce, technology and national wealth,
Sprat was the heir to ideas that had been championed by Hartlib's
friends in the Interregnum. In both cases an appeal was made to the
national good in an attempt to propitiate all.[76]

Sprat's *History* thus gives a full view of the aspirations of Restoration science, and it is a very valuable document. But it is wrong to
mistake ambitions for achievement, and his book has often been
misunderstood for this reason. Not all the hopes for science that he
expressed or the philosophical positions that he espoused proved as
widely acceptable as he expected, as subsequent controversy about
the religious and intellectual implications of his work revealed.[77] The
remaining chapters of this book will therefore examine the validity
of Sprat's views. How far was science supported by the classes to
whom Sprat hoped it would appeal? How close were the links
between science and economic life that he aspired to in his rhetoric
about 'a *Philosophy*, for the use of *Cities*, and not for the retirements
of *Schools*'?[78] Indeed, how legitimate was it virtually to identify
Restoration science with the Royal Society, as Sprat did? To this we
must now turn.

---

[75] Sprat,[92] p. 78 and passim.
[76] For comparable statements by others see Jacob[88]; on the Hartlibian background,
Webster,[72] esp. pp. 446–65.
[77] See below, ch. 6.   [78] Sprat,[92] p. 76.

2

The significance of the Royal Society

The Royal Society was founded in the year of Charles II's Restoration and the inauguration of this first scientific institution undoubtedly distinguishes science after 1660 from its Interregnum precursors. This, however, has sometimes led to a neglect of these predecessors and a presumption that the Society achieved more than it really did. The Society's subsequent continuity to become the country's foremost scientific society in the twentieth century invites anachronistic assessments of its early endeavours which it is essential to avoid. A correct approach to the Society gives much insight into Restoration science which incorrect analysis obscures.

The early Royal Society was undeniably significant in what might be called its 'definitional' capacity, which mattered as much in its own time as since. It was specifically devoted to the study of natural philosophy – to what was emerging as 'science' in a recognisably modern sense – centring on natural and mechanical problems but extending through the life sciences towards medicine and through chemistry and applied mathematics towards technology. Hence both hostile and favourable contemporaries used the Society's name as shorthand for the hopes and achievement of the new science as a whole. Glanvill in *Plus Ultra* and Sprat in his *History of the Royal Society* conflated the Society's work with all of recent progress in its chosen fields, and this tendency to coalesce the Society with Restoration science has continued ever since.

It is far from clear, however, whether this was primarily a matter of convenience – the Society being a mere symbol for developments it had little to do with – or whether these stemmed from new circumstances for which it deserves the credit. This chapter seeks to throw light on the problem from the circumstances of the Society's incorporation, its expressed ambitions and its actual history. Only by thus illustrating the advantages and disadvantages of the impulse to institutionalise science in this formative period can one avoid the vapid and untestable generalisation into which numerous commentators on the Society have fallen.

The Royal Society has often been acclaimed as the chief inspiration of Restoration science. The most recent full-length study – Margery Purver's *The Royal Society: concept and creation* (1967) – has only taken a commonplace to extremes in arguing for the Society's decisive influence in its early years through its articulate and active championship of a sophisticated Baconian programme of co-operative induction. This, it is said, assured progress in natural knowledge where other, less well-organised previous attempts had failed, and thus explains the scientific advances of the late seventeenth century. Miss Purver's view, however, has rightly been severely criticised. It is undoubtedly simple-minded to postulate a straightforward Baconian orthodoxy in late seventeenth-century England, or to see this as the only key to intellectual advance. Most of the contrasts and conflicts in method and theory surveyed in chapter 1 differentiate Fellows of the Royal Society, and it would be surprising if the Society had introduced a new orthodoxy overnight. It has even been suggested that the Society made so much of its adherence to Bacon's programme at least partly to disguise the actual diversity of its members' interests and techniques.[1]

Indeed critics have argued that it is wrong to see the Society's early history as a success story largely because of the major scientific achievements of the age with which the Society could claim a formal association. If its aspirations as a corporate entity are considered instead of the sum of its Fellows' individual accomplishments, its deficiencies as a body come into prominence. Attention has been drawn to the Society's 'institutional weakness' and its 'rapid decline', views which (as we shall see) have much to support them, and the juxtaposition of scientific vitality with the existence of the Society might seem almost coincidental.[2]

Perhaps most extreme is the view of the Society not as 'the conscious centre of all genuinely scientific endeavour, but something much more like a gentlemen's club', and this too is not without some truth.[3] The Society was the heir of the Interregnum scientific groups in London and Oxford, surveyed in the last chapter, and they had a strong social element, meeting in coffee-houses and private

---

[1] Webster,[102] pp. 114f., 125.   [2] Webster,[205] p. 22.   [3] Skinner,[109] p. 238.

lodgings.[4] The relative informality of these gatherings partly explains why the Society's origins have been so controversial, since scholars have often looked for more precisely defined coteries than actually existed. Even after 1660, the Society continued to thrive in the context of 'clubbing' whose growth is characteristic of seventeenth-century London social life. Meetings were followed by dinner parties marked by 'good chear, & company', while writings that Marcello Malpighi sent to the Society from Bologna in 1678 – presumably the second volume of his *Anatome Plantarum*, published the following year – were discussed in a coffee-house by Hooke and his friends before being formally presented at a meeting.[5] From the first the Society gained a 'social' intake (as the Oxford group had previously) of people vaguely interested in science but not actively concerned with it, that component of fashionable curiosity about the new science classically sketched by Lord Macaulay in his *History of England* which will be examined in chapter 3. Contemporaries were well aware that a significant portion of the Society came to meetings 'only as to a Play to amuse themselves for an hour or so'.[6]

The Royal Society was also a London club in that its foundation was precipitated by the effect of the Restoration in bringing to the metropolis many men who had been scattered elsewhere during the 1650s – some in Oxford, others abroad – to take advantage of the opportunities offered by the new regime. These were glad to congregate to share their scientific interests with each other and with those who had been in London throughout the Interregnum: those who came from Oxford were accustomed to meeting there and one author has seen the Society's foundation as an attempt by university men to continue a research atmosphere in a metropolitan environment.[7] So it was a club for Londoners, and though it comprised the largest single body of scientific enthusiasts in the country, this limited its usefulness to those outside the capital. In fact it was impossible for these to take part in the Society's activities except by correspondence, and many famous scientists whose names appear in

---

[4] Wood, *Life and Times*, I, 201; see also ibid, I, 472–3; Webster,[72] p. 155.

[5] Brooke to Lister, 14 Dec. 1672, Lister 34, fol. 76; cf. Pepys,[163] VI, 36; Hooke,[165] p. 387.

[6] Hartley,[97] p. 13; Hunter,[107] pp. 13–14 and passim; T. B. Macaulay, *The history of England*, I (London, 1849), 405–7; DM 5, fol. 12.

[7] Frank,[132] p. 199. On the circumstances of the Society's foundation see especially Webster,[72] pp. 88–99.

the list of members had very little to do with the Society because their place of residence made this physically impossible.[8]

Yet against this one has to balance much larger ambitions for the Society. Its background was a widespread enthusiasm for institutionalisation as a more productive way of pursuing scientific enterprise. The Society's closest precedent is to be sought in the rules drawn up for the Oxford experimental philosophers in 1651, which mention regular meetings, admittance by majority vote, a 'Catalogue' of members and the levy of subscriptions to pay for publicly exhibited experiments.[9] More ambitious plans were definitely 'in the air' around 1660, as is shown by the proliferation of proposals for institutions devoted to intellectual activities, including Abraham Cowley's 'Philosophical Colledge' and the Swede, Bengt Skytte's, utopian community.[10]

The stimuli behind these and the Royal Society were diverse, but all manifest a feeling that the advancement of learning was best forwarded by organisation. One inspiration was clearly Bacon's vision of 'Solomon's House' in his *New Atlantis*, and the Society may also have owed something to continental utopias like J. V. Andreae's *Christianopolis* (1619).[11] The example of comparable institutions in France and Italy also played its part. The founders of the Royal Society alluded to these at their inaugural meeting on 28 November 1660, hoping to initiate 'a more regular way of debating things; and that, according to the manner in other countries, where there were voluntary associations of men into academies for advancement of various parts of learning they might do something answerable here for the promoting of experimental philosophy'. This is a clear allusion to the *académies privées* in France, while another forerunner – though any direct influence is hard to prove – was the Accademia del Cimento in Florence, perhaps the Royal Society's nearest precursor in its programme of co-operative experiment.[12]

Moreover no aspect of the Royal Society in its early years is more striking than the formality of its aspirations and activities, part of its ambition to be a real national institution focusing scientific enterprise rather than a mere intellectual social club. Most important was

---

[8] Hunter,[107] pp. 11–12, 58–60.   [9] Hartley,[97] pp. 25–6.
[10] Cowley, *A Proposition For the Advancement of Experimental Philosophy* (London, 1661); Purver,[24] pp. 220–1, 227–32; Evelyn to Boyle, 3 Sept. 1659, Evelyn,[126] pp. 116–20.
[11] F. E. Held, *Christianopolis: an ideal state of the seventeenth century* (New York, 1916), ch. 5.   [12] Birch,[118] I, 3; Brown[101]; Middleton,[100] ch. 6 and passim.

the granting of a royal charter in 1662 which was replaced by a second the following year: this made the body far more permanent and formal, for it now joined the ranks of the chartered corporations, with a constitution which had to be observed and which conferred rights and privileges.[13] Not the least of these – and indicative of the more public role that the charter gave – was the permission conferred on the Society's officers to license books under the 1662 Licensing Act. The use of this 'Imprimatur' by the Society's officers instead of the bishops otherwise responsible for approving books on science and philosophy might have been crucial had the Society used it boldly at a time when the censors were notoriously cautious, though in the event it was employed rather soberly.[14]

Institutionalisation also involved a constitutional structure, providing for the election of officers, for minimum quora at such elections and for the ownership of corporate funds and possessions. The conscious formality of the Society's early years was perhaps most evident in the way that meetings were conducted, on which foreign observers remarked. No meeting could be held unless the President or one of the Vice-Presidents was in attendance, nor in the absence of a quorum, and proceedings were regulated by the chairman with a mallet. Most telling was the requirement that the President alone should wear his hat while all others doffed theirs when addressing him, a strange gesture to the superior role of the chairman redolent of the social structure of traditional England.[15] In addition, the greater permanence ensured by the constitution encouraged the careful preservation of minutes and of papers read at meetings. This gave continuity and provided an invaluable record, the absence of which had been seen as a drawback of the Oxford group.[16]

More important, a conscious effort was made to enrol the most eminent practitioners of subjects at the heart of the Society's programme as members, whether or not they were frequently in London and hence able to attend meetings. Thus the membership list – which was printed and issued annually – at least appeared to be truly representative of national achievement in such fields, even if some

13 Birch,[118] I, 88–96, 221–30.

14 For a list of the books published thus see *The Record of the Royal Society*, fourth edition (London, 1940), pp. 36–8; it is not, however, completely accurate.

15 Hunter,[107] pp. 16, 19–20; Samuel Sorbière, *Relation d'un Voyage en Angleterre* (Paris, 1664), pp. 87, 88–91; Balthazar de Monconys, *Iournal des Voyages* (Lyons, 1666), II, 26; Birch,[118] I, 264, II, 95, 193; Lawrence Stone, *The crisis of the aristocracy* (Oxford, 1965), pp. 34–5.　　　　　16 Hooke,[65] p. iii.

were only nominal members. In the Society's earliest years the Cambridge Platonists, Henry More and Ralph Cudworth, were both proposed for membership and More's name appeared in the Society's lists for the rest of his life although he barely took part in its activities; others in this category include Isaac Barrow in Cambridge and James Gregory in Scotland.[17]

More important than such rather symbolic gestures was the Society's wish to play a corporate role in the advancement of learning. This can be reconstructed from numerous more or less public statements by those associated with the Society in its early years, particularly the letters of the first Secretary, Henry Oldenburg, and Sprat's *History of the Royal Society*. The Society first and foremost espoused a Baconian programme of experiment and record. As Oldenburg put it, 'It is our business, in the first place, to scrutinize the whole of Nature and to investigate its activity and powers by means of observations and experiments; and then in course of time to hammer out a more solid philosophy and more ample amenities of civilization.'[18] To this end various steps were taken, particularly towards collecting information, some of them echoing projects suggested earlier at Oxford. Grandiose plans were made to scrutinise all previous work on related subjects, to collect 'all philosophicall experiments hitherto observed made and recorded' and to see which matters needed further investigation. Perhaps the Society's most characteristic occupation in its earliest years was drafting questionnaires relating to particular topics and geographical areas about which further facts were required.[19] In 1664 a number of specialised committees were formed to orchestrate this large-scale collaborative activity in all parts of natural knowledge, and the very scale of the enterprise is impressive. As Oldenburg told John Winthrop, Governor of Connecticut, in 1667: 'Sir, You will please to remember, that we have taken to taske the whole Universe, and that we were obliged to doe so by the nature of our dessein.'[20]

But merely collecting information was only one of the Society's ambitions. It was, after all, particularly devoted to 'experimental philosophy', and it is hence important not to underestimate its initial

[17] Hunter,[107] FF114 (though cf. Pepys,[163] VIII, 243), 116, 115, 264.
[18] Oldenburg to van Dam, 23 Jan. 1663, Oldenburg,[121] II, 13–14.
[19] H. W. Robinson, 'An unpublished letter of Dr Seth Ward relating to the early meetings of the Oxford Philosophical Society', *N.&.R.R.S.*, 7 (1949), 69; DM 5, fol. 62; Cl.P. XIX, passim.
[20] Birch,[118] I, 406–7; Oldenburg to Winthrop, 13 Oct. 1667, Oldenburg,[121] III, 525.

commitment to corporate experiment. For though the Royal Society might have been expected to be primarily a forum where ideas were reported and discussed and no actual scientific work done, in the manner of a modern learned society, in fact the contrary was the case. Its founders were convinced that the Society was itself the proper place for experiments, and high value was placed on the consensus to be achieved in evaluating their results with many men witnessing them. Sprat stressed how experiments were tried 'till the whole *Company* has been fully satisfi'd of the certainty and constancy; or, on the other side, of the absolute impossibility of the effect'.[21] The Society was thus not just a clearing house for information but an active centre of scientific research. As such, it was typical of the strong experimental impulse in the science of the time, seen also in the Accademia del Cimento or in the Académie des Sciences, founded in Paris in 1666. Both were rather less 'public' than the early Royal Society, but both saw the corporate pursuit of an experimental policy as the most satisfactory way to advance scientific knowledge.[22]

More than this, the Society hankered after the status of a national research institute with a salaried staff of expert scientists in custom-built and properly-equipped premises, all this paid for by large-scale endowment, perhaps from public sources. In the early 1660s, especially, there were expectations that the Society might be handsomely endowed. Fellows expended considerable time and ingenuity on ways of raising money for the Society, if not by using courtly contacts to gain a direct grant from the King, then by thinking of some privilege which he might give which would benefit the Society without penalising anyone else. The possibilities considered ranged from bestowing salaried posts on its officers for the Society's benefit to reviving old monopolies on its behalf.[23] In addition to the King, it was hoped that funds might be forthcoming from the gentry and nobility to whom the ideology outlined by Sprat was supposed to appeal. One request for money to the landed classes of Ireland survives, which characteristically stresses the utility of the Society's design not only in improving natural knowledge by checking things hitherto taken on trust 'as also for perfecting all usefull mechanicall uses & practices' (an addition to the draft). It may be significant that this was aimed at the cockpit of earlier Hartlibian attention and it

---

[21] Frank,[106] pp. 83–4; Sprat,[92] p. 99.   [22] Middleton;[100] Hahn,[99] ch. 1.
[23] Birch,[118] I, 377, 379; Hunter,[107] pp. 16–17; DM 5, fol. 37.

provides an element of continuity between the Hartlib circle and the Royal Society.[24]

Similar ambitions underlay the plan devised in the late 1660s for building and endowing a 'college' for the Society, for which subscriptions were solicited both from Fellows and from 'other well disposed persons not of the Society'. This was to have provided a permanent base for the Society, 'where we may meet, prepare and make our Experiments and Observations, lodge our Curators and Operators, have our Laboratory, Observatory and Operatory all together', as Oldenburg put it, seeing it as 'a means very probable to establish their Institution to perpetuity'. Similar sentiments were echoed on the form of subscription proffered to those who it was hoped might contribute, 'being satisfied of the great usefulness of the institution of the Royal Society, and how requisite it is for attaining the ends designed thereby, to build a college for their meetings, and to establish some revenue for discharging the expenses necessary for trial of experiments'.[25]

It was aspirations of this kind – to become something very much more than a voluntary group of enthusiasts – which apparently did most to cause friction between the Society and other learned bodies with comparable ends, the universities and the College of Physicians. If entrenched in this way, the Royal Society might have become a new kind of national, metropolitan institution, devoted to the systematic reform of knowledge along lines hitherto neglected, and the sheer grandeur of such objectives is the background to the rumours of empire-building in related studies which raised the hackles of potential rivals.[26]

The first casualty among these grandiose plans was the hope for large-scale endowment and proper 'establishment'. Even the college came to nothing, although a number of subscriptions were promised, a site earmarked and more than one design for the building prepared.[27] Instead, throughout this period, the Society had to base itself in other premises, for the most part in the declining Gresham College, founded to promote education in the City in the Elizabethan period but now on hard times. The Society worked out a sort of symbiotic relationship with Gresham from which both parties gained, the Society obtaining a spacious rent-free home while the college derived some desperately needed vitality. After the Fire of

[24] DM 5, fol. 34v. [25] Hunter,[107] pp. 60–2; Birch,[118] II, 205, 211.
[26] See below, ch. 6. [27] Hunter,[107] p. 61.

London, Gresham was temporarily commandeered for civic functions and an alternative home was provided by the Duke of Norfolk
in his mansion, Arundel House, in the Strand. But the Society returned to Gresham in 1673 and remained there until the early
eighteenth century.[28]

Similarly, the King's response (like that of the Irish landowners) to
the Society's requests for endowment was intensely disappointing.
Such schemes came nearest to fruition in a royal grant of fractions
accruing from the Restoration Irish land settlement in 1663, but
unfortunately there were rival claimants, so nothing came of this for
the Society. The King did grant the Society Chelsea College, founded
by James I as a Protestant seminary but now defunct. This too,
however, proved something of a white elephant, involving members
in years of legal wrangling and unrewarding upkeep. Only in 1682,
when the King bought it back from the Society in circumstances that
will be examined in chapter 5, did the Society obtain a sizeable
regular income for the first time, and by that date its development
was already firmly fixed along lines determined in the two earlier
decades.[29]

This lack of endowment was a matter of constant regret to at least
some of the Society's organisers. Oldenburg's letters are full of sad
reflections on what the Society might have achieved had it ever had
proper endowment: 'no question, this Society would prove a mighty
and important Body, if they had but any competent stock to carry on
their desseins'.[30] Oldenburg and others deeply envied the Académie
des Sciences established by Louis XIV, encouraged by his great
minister Colbert, with salaries for research-workers and lavish facilities. This seemed likely to assure scientific advance stunted by the
impoverishment of natural philosophy in England. As John Evelyn
complained, 'We see how greedily the *French*, and other *Strangers*
embrace and cultivate the *design*: what sumptuous *Buildings*, well
furnish'd *Observatories*, ample *Appointments*, *Salaries*, and *Acco-
modations* they have erected to carry on the *Work*; whilst we live
Precariously, and spin the *Web* out of our own *bowels*.'[31]

Instead, the English society remained very much an amateur body
– 'a great assembly of Gentlemen', as a contemporary put it – almost

[28] Adamson,[110] pp. 1–7; Adamson,[111] ch. 7.   [29] Hunter,[107] p. 17; DM 5, fol. 19.
[30] Oldenburg to Boyle, 24 Nov. 1664, Oldenburg,[121] II, 320.
[31] Oldenburg to Williamson, 4 July 1668; Oldenburg,[121] IV, 498–9; Oldenburg to
    Hevelius, 12 June 1671, ibid, VIII, 97–8; Evelyn,[197] sig. A3v.

entirely dependent for funds on its Fellows' subscriptions, a mere two guineas a year per member. This, even at the best of times, gave an annual revenue of only a couple of hundred pounds, though extra contributions could occasionally be raised; at its peak the membership reached 228, but many Fellows were irregular in paying their dues.[32] It was claimed that financial strain 'subiected the Society to promiscuous, & lesse-usefull Members, & Elections, which have Eclips'd its Reputation', and the hope of money undoubtedly made the Society's organisers reluctant to stem the recruitment of more or less interested amateurs, who were sometimes seen as an embarrassment. Perhaps the main change of policy made possible by the ultimate endowment of the Society in 1682 were the sizeable expulsions that now occurred for the first time, in contrast to the greater hesitance shown in the 1670s.[33]

In the event the Society's achievement in this direction was not negligible considering its limited resources, and funds from members were judiciously used to pay professionals to do research and carry on the Society's business. Oldenburg received a salary from the Society at least from 1669; he possibly had financial support earlier to supplement his haphazard earnings from other sources.[34] The Society may also claim some credit for maintaining Robert Hooke, who was paid a small salary for his assiduous work as Curator of Experiments, though he also received money from Gresham College and the City of London. More important, in 1672 Nehemiah Grew was lured from Coventry to London by the Society's offer of a stipend which was raised by special subscription and which enabled him to complete and publish his pioneering work on plant anatomy. At a humbler level, in 1669 the Society engaged the self-educated naturalist, Thomas Willisel, to collect botanical and zoological specimens around the British Isles.[35]

Plans in the 1660s to establish an anatomy lectureship for such medical Fellows of the Society as Walter Charleton, Richard Lower or Walter Needham, on the other hand, did not materialise, and the

---

[32] Hunter,[107] pp. 10, 17–18; Bluhm,[108] p. 83.

[33] Evelyn to Colwall, 11 Feb. 1682, Evelyn, *Lb.*, no. 439; Hoppen,[117] pp. 86–7; Birch,[118] IV, 157–8; Hunter,[107] pp. 54–7 (on one small earlier bequest, see ibid, n. 42).

[34] Birch,[118] II, 273, 376; Oldenburg,[121] II, xxiv–vi, IV, xxiv–v, V, xxv–vi, VI, xxviii.

[35] 'Espinasse,[136] pp. 4, 83; Birch,[118] III, 42, 69–70; Hunter,[107] p. 17; Jeanne Bolam, 'The botanical works of Nehemiah Grew, F.R.S.', *N.&.R.R.S.*, 27 (1973), 219–31; Birch,[118] II, 358, 371, 378–9.

Society's shortage of money often thwarted its ambitions. The only publication in which it had a direct financial interest – Francis Willughby's *Historia Piscium* (1686) – proved an embarrassing financial burden which the Society could ill afford.[36] Even its facilities depended mainly on gifts from generous Fellows. The Society's 'repository' or museum will be considered in chapter 3, and it also had some success in accumulating a library of scientific and mathematical books and a collection of microscopes, telescopes and other equipment. But the Society fell far short of its wish to provide a facility not otherwise available in London: both its equipment and its books remained inferior to the personal facilities of certain Fellows, as is revealed by the sale of Hooke's magnificent library of over 3000 volumes after his death.[37]

The failure to realise this part of the Society's aspirations did not mean, however, that the more general hopes for its corporate activity were equally doomed. In fact the plans for a college revealed a split within the membership between those bitterly aware of how much more might have been achieved with better facilities and those perfectly happy with the Society as it was. Some members may have shared the reservations expressed by Christiaan Huygens in France as to whether it was wise for science to become as dependent on 'the Humour of a Prince & the favour of a minister' as was the Académie des Sciences.[38] Sprat indicated that his programme for co-operative research was by no means dependent on endowment: he argued that a start should be made with purely voluntary finance and that fuller funding should only be expected when the Society had proved its value. Clearly many members thought that the reform of knowledge could begin through what has aptly been described as 'a kind of once-a-week Rockefeller Institute'.[39]

But this too proved difficult. Though corporate experiment was intended to be the staple of meetings, the Society's profuse minutes reveal a steady decline in experimental trials over its first two decades. Instead, proceedings were increasingly taken up with reports on work done elsewhere, discussions of this and its significance, and

[36] Birch,[118] I, 415, II, 178–9, 206, 212, 355; Bluhm,[108] pp. 98–100.
[37] Hooke to Bernard, 7 April 1674, Smith 45, fol. 105; Nehemiah Grew, *Musæum Regalis Societatis* (London, 1681), pp. 357–60 and passim; H. A. Feisenberger, ed., *Sale catalogues of libraries of eminent persons*, vol. XI (London, 1975), pp. 37–116, reprints the auction catalogue of Hooke's library.
[38] Hunter,[107] p. 61; Vernon to Oldenburg, 15 Feb. 1670, Oldenburg,[121] VI, 505.
[39] Sprat,[92] pp. 77–80; Frank,[106] p. 84.

the exhibition of objects. The Society thus unconsciously evolved into something much more similar to a modern scientific society in which report and debate rather than actual scientific inquiry dominated the business. Nor did another of the Society's intended corporate activities, the specialised committees set up in 1664, prove very successful: there were soon complaints that they 'fall to the ground, because tis not possible, to bring people together', and they rapidly succumbed entirely.[40]

To some extent the decline of collaborative experiment was due to the innate impracticality of the Society's original stress on co-operative scientific research. The same dwindling of corporate experiment recurs in the Académie des Sciences, despite its differences from the Royal Society, and this reflects an institutional embodiment of the conflict between Baconian and theoretical science surveyed in chapter 1: experiment and record were only part of the mechanism of scientific progress, as was thus painfully discovered.[41] At the Royal Society, however, the problem was enhanced by the nature of the organisation, showing (as a recent study has indicated) the difficulties of trying 'to carry out scientific activity which is institutionalised, but not professionalised'.[42] It also illustrates a genuine tension between the grandiose plans to reform knowledge by organised intellectual activity and the strong social element present in the Society from its beginnings.

Quite apart from a tendency for some subjects to predominate among the experiments because they were easily demonstrated, the proceedings were prone to develop along the lines of a gentlemen's club. Statistical study reveals that those of high rank were likelier to contribute to discussion than their humble and diffident colleagues, while, whatever the actual attendance, verbal participation was apt to fall to a few people. The turnover of these was surprisingly high, so that the consideration of topics lacked continuity and was often amateurish, since, apart from a few old stalwarts, the most active Fellows were frequently the newest. In addition, the kind of scientific activity likeliest to appeal to a public audience tended to predominate, not least because the Society's institutional structure set a premium on a large rather than a well-informed audience, and, conversely, this put serious scientists off coming to meetings.[43]

---

[40] Frank,[106] pp. 94f; Oldenburg to Boyle, 22 Sept. 1664, Oldenburg,[121] II, 235; Lennard,[179] pp. 27–8.   [41] Hahn,[99] pp. 24–30.   [42] Frank,[106] p. 82.
[43] Frank,[106] pp. 87, 92–4, 96–7; Hunter,[107] pp. 20, 23–32.

John Evelyn once disarmingly described the Royal Society as 'an Assembly of many honorable Gentlemen, who meete inoffensively together under his Majesty's Royal Cognizance; and to entertaine themselves ingenuously, whilst their other domestique avocations or publique businesse deprives them of being allwayes in the company of learned men and that they cannot dwell forever in the Universities'.[44] This is a far cry from the strident claims of Sprat and others, but it makes better sense of the often rather miscellaneous and trivial discussions recorded in the Society's minutes. If the Society's proceedings were sometimes dominated by a merely fashionable audience, however, this was fickle in any case (as a portion of the audience for science in Interregnum Oxford had been earlier). Though the Society served a valuable function by bringing together major and minor scientific figures – 'the chiefe advantage of Royal Society membership', in the view of one early Fellow, Sir John Hoskyns – there were other places for this, such as the London coffee-houses. Hence there was a steady tendency for lesser men to fall away, and the Society had repeatedly to prune dead wood from its ranks.[45]

Matters were exacerbated by severe internal tensions between powerful personalities. The Society seems to have suffered more from these in its early years than most such bodies, and quarrels certainly hampered its activity. The President for the first fifteen years of the Society's chartered existence, William, Viscount Brouncker, was a difficult man who made many enemies. In a letter to John Aubrey, Walter Charleton scurrilously remarked that it was impossible for Brouncker 'to oblige any but a Miss', and he advised Aubrey:

if you would make him your Patron & raiser, you have no other way to doe it, but by bribing his mercenary [i.e., his mistress] – who by that means alone became his, after she had passd through almost as many hands, as the Royal Society hath members, & many more than she has teeth in her gumms of Natures setting.[46]

From the 1670s there was also strong personal antagonism between the Society's two principal paid servants, Hooke and Oldenburg. This was brought to a head by Oldenburg's attempt to secure the English patent for Huygens's spring-balance watch, which (charac-

---

[44] Evelyn to Beale, 27 July 1670, Evelyn, *Lb.*, no. 329.
[45] Hartley,[97] p. 13; Hoskyns to Aubrey, 20 Feb. 1675, Aubrey 12, fol. 215; Hunter,[107] pp. 23–32 and passim.
[46] Charleton to Aubrey, 4 Feb. 1672, Aubrey 12, fol. 66.

teristically) Hooke claimed had been plagiarised from him. Brouncker sided with Oldenburg, and the quarrel culminated in Hooke's setting up a splinter group, a 'select clubb', which, after starting as a slightly more formal version of his personal coffee-house clique, expanded to include a cross-section of the Society who were tired of the Brouncker–Oldenburg regime. This is evidenced by Hooke's *Diary*, which is generally an informative source on personal tensions in scientific circles.[47]

Such problems continued in the 1680s. Martin Lister withdrew from the Society after being active in it for a few years and he was later vociferous in complaining about its tendency to cliquishness: 'such meane & invidious spirits reigns amongst even Societies found-ed purposlie for the promoting of Learning in all its parts'.[48] Ten-sions continued in the 1690s, when John Evelyn complained about the acrimony between Tancred Robinson, John Woodward and other Fellows, 'Methinks, Philosophers should not fall-out about shells & pibbles.' By 1700 these quarrels had reached such intensity that there were fears that they might 'end in the destruction of the Royal Society'.[49] As a result, once-active members were continually prone to withdraw. Careful study of eminent scientists who joined the Society before 1685 reveals that, after an initial period of enthu-siasm, many ceased to attend the Society's meetings with any reg-ularity although they continued to prosecute related interests: the list includes Walter Charleton, Richard Lower, Seth Ward, Christopher Merrett, Sir George Ent and Walter Needham.[50]

It is therefore hardly surprising that the Society suffered repeated crises in its London institutional life as numbers and activities dwin-dled and had to be forcibly and sometimes artificially revived. Even in the 1660s Oldenburg's letters to Boyle sometimes complain about the Society's problems; in 1671 a country correspondent reported that Oldenburg 'seemed to be in an agony for feare of an eclipse of the Royal Society'; and in 1673–5 a major programme of reinvigora-tion was implemented, including the soliciting and publishing of research papers, an energetic attack on the mounting arrears in subscriptions, and the sloughing off of unwanted members. This led

---

[47] 'Espinasse,[136] pp. 63–70; Hooke to Aubrey, 24 Aug. 1675, Aubrey 12, fol. 186; Hooke,[165] esp. pp. 199–200, 205, 239.

[48] Carr,[142] pp. 40–1; Lister to Lhwyd, 6 Feb. 1696, Bodleian Library MS Ashmole 1816, fol. 124v.

[49] Evelyn to Tenison, 26 Sept. 1697, Evelyn, *Lb.*, no. 794; Robinson to Lister, 26 March 1700, Lister 37, fol. 26.   [50] Hunter,[107] FF 68, 234, 29, 17, 31, 230.

to a temporary improvement, but by 1680 matters seem to have been almost equally bad, and the Society decided on a policy of Baconian retrenchment.[51] Earlier anxieties were echoed in the late 1680s, when Tancred Robinson reported to Hans Sloane, then in Jamaica: 'the Royall Society declines apace, not one Correspondent in being . . . The same officers are chosen this year: I am afraid that you will find nothing but ruines at your return, all will appear an Alsace to you.' Sloane's own Secretaryship in the 1690s saw some improvement, but in 1699 Sir Robert Southwell, another active Fellow, reported that 'the Royal Society's Stock is so low as hardly to keep Life and Soul together'.[52]

It thus cannot be wondered at that – whatever its ambitions – the Royal Society was never the chief centre of research even for men with scientific interests who lived in London. The Society's most dedicated and loyal Fellows actually 'did' most of their science on their own, including Boyle (living in London from 1668), whose 'ample and well furnished' laboratory in his house in Pall Mall was widely admired.[53] There were also rival foci of scientific discussion, including the College of Physicians, which continued its Interregnum role at least in part. In the 1690s naturalists like Lister could find an alternative meeting-place at the Temple Coffee House, where various Fellows of the Royal Society and others met to talk about their mutual concerns.[54]

Furthermore, quite apart from the Fellows who drifted away, not all those elected and included in the printed lists of Fellows had anything to do with the Society, though they often actively pursued scientific interests elsewhere. Thus Thomas Willis, despite his close association with the group of natural philosophers in Interregnum Oxford, had hardly anything to do with the Society after moving to London. Neither did the immigrant mathematician, Nicholas Mercator, in spite of his connections with many Fellows.[55]

If this was true of Londoners for whom the Society might have been a natural focus, it is not surprising that it was all the more so with scientists elsewhere, who, even when elected to the Society, were obliged to work separately. One such was Henry Power in

[51] Hunter,[107] pp. 24–32; Evelyn to Pepys, 25 June 1680, *H.M.C. MSS of J. Eliot Hodgkin* (1897), p. 177; Evelyn,[164] IV, 205.
[52] Robinson to Sloane, 6 Dec. 1687, Sloane 4036, fol. 30; Levine,[261] p. 85.
[53] Maddison,[140] pp. 136, 144, 187.
[54] See below, p. 144; Carr,[142] p. 41; Stearns,[151] pp. 253–4; Allen,[317] pp. 10–11.
[55] Hunter,[107] FF 90, 215.

Halifax, who was briefly associated with the Society in the early 1660s. Power's scientific output, however, was largely independent of it even in these years, and he evidently got more from his contacts with the circle of Richard Towneley in Lancashire.[56] John Ray, also, had relatively little to do with the Society's corporate activities though he reported on his work by letter. Neither did the two greatest scientists of Restoration Cambridge, Isaac Barrow and Isaac Newton, both of whom held professorships there and worked mainly on their own despite their membership of the Society.[57]

Indeed both in London and the provinces some with achievements in science to their credit never even applied for membership. In London, these included such doctors as Thomas Sydenham and Richard Wiseman, whose interests overlapped with those of the Society and who were friendly with its Fellows. Further afield, there was Richard Towneley in Lancashire and Robert Morison, Professor of Botany at Oxford, both of whom were nonetheless on perfectly good terms with the Society.[58] Though, as we saw earlier, the Society made some attempt to enrol those eminent in natural philosophy regardless of their lack of involvement in the Society's meetings, this policy was not so systematic that excuses have to be made for a scientist's absence from the Society's ranks. It is anachronistic to see the Society as a prototype of an exclusive twentieth-century professional institution.

It is equally wrong to search for a single Royal Society orthodoxy, shared by all who belonged and negated by those who did not. Towneley, it is true, espoused a rigid Cartesianism which Boyle and other prominent members found distasteful, and it is interesting that Power's contact with the Royal Society in the early 1660s temporarily lent his work an untypically empirical tone. But too much should not be made of this. Equally telling, a non-member like Sydenham combined a vocal inductivism with an unacknowledged element of theory that is very similar to the Baconianism of many Fellows of the Society: yet the Society also contained men with deductivist attitudes like Power's.[59] The Royal Society was representative of the general character of Restoration science, but this was largely due to random selection.

[56] Webster,[158] pp. 66–8.   [57] Hunter,[107] FF 239, 115, 290.
[58] Skinner,[109] pp. 236–7; Kenneth Dewhurst, *Dr Thomas Sydenham* (London, 1966), esp. pp. 63–4; *Phil. Trans.*, 8 (1673), 6052–4; Webster,[158] pp. 62–74.
[59] Webster,[29] pp. 167–8; Wood,[28] pp. 172–4; King,[41] pp. 113–33.

This estimate may seem to conflict with the strikingly high esteem in which the Royal Society was held throughout the Restoration, among scientists and non-scientists, at home and abroad. This appears not least in the ecstatic opinions which fill foreigners' letters to the Secretary, Oldenburg: they saw the Society as England's 'chiefest ornament', claiming that its Fellows 'need not envy' Aristotle's Lyceum, Plato's Academy or Zeno's Stoa.[60] Such was the Society's foreign reputation that Oldenburg complained of being plagued by continental visitors to London who prevented him from getting on with his affairs.[61] Adulation of the Society can be found in England, too, not least in the provinces: thus John Burton, a country schoolmaster, protested that he 'had no other designe, nor higher ambition, than to express my service and zeale to the illustrious Society'.[62]

The Society undoubtedly gained respect as a royal foundation and as the institutional embodiment of the new science, regardless of what it actually did. Non-Fellows with related interests wrote fulsome prefaces to their books in which they praised Charles II for founding this '*Royal and most Illustrious Society*'. In his *Metallographia* (1671), John Webster saw it as 'one of the happy fruits of His Majesties blessed and miraculous Restauration, and that which will speak him glorious to all succeeding Generations, beyond all his Royal Progenitors'. The Society's name provided a useful symbol of the new science which itself harmonised with the almost jingoistic royalism of the Restoration.[63]

The Society certainly deserves credit for enhancing science's social respectability by recruiting eminent public figures to its ranks. Its membership lists, profusely decorated with the names of (mainly inactive) bishops, statesmen and aristocrats, enjoyed wide circulation, disseminating esteem for the Society among non-members at home and abroad. The lists obviously gave the impression that the new science had been espoused by the establishment to an extent that was not necessarily the case, and the Society retained its august aura long after the time (mainly in the late 1660s) when the bulk of such

---

[60] Sterpin to Oldenburg, 24 Feb. 1671, Oldenburg,[121] VII, 469; Nazari to Oldenburg, 5 Oct. 1669, ibid, VI, 258.

[61] Oldenburg to Boyle, 30 Mar. 1668, Oldenburg,[121] IV, 282.

[62] Burton to Oldenburg, 9 Dec. 1667, Oldenburg,[121] IV, 21.

[63] John Worlidge, *Systema Agriculturae* (London, 1669), sig. D2; John Webster, *Metallographia* (London, 1671), sig. [a]2v; cf. Samuel Sturmy, *The Mariners Magazine* (London, 1669), sig. A2v.

recruitment occurred, not to be repeated on this scale in later years. A related function was doubtless served by the emphasis in the Society's publications on the noble Robert Boyle, a member of a distinguished aristocratic family who was closely connected with the establishment.[64]

If the Society had some undeserved reputation outside its London base, however, much of its fame was merited. For the Society met a real need in organising the science of the day – in gathering information and stimulating, co-ordinating and publicising the research of scientists who, despite the Society, continued to work on their own. This capacity had been only partially foreseen by the Society's founders, but it went far towards transcending the London meetings as the Society's chief claim to importance in the history of science. It was the achievement mainly of one man, Henry Oldenburg, a German emigré who had formerly been tutor to the Boyle family and who held the Secretaryship of the Royal Society from 1662 until his death in 1677.

Oldenburg worked prolifically on the Society's behalf, corresponding with major and minor scientists at home and abroad, and his correspondence, which is now being published, is one of the most important sources for the history of the science of his day. To some extent Oldenburg's role continued that of earlier intelligencers like Marin Mersenne in France and Samuel Hartlib in Interregnum England. Hartlib had been convinced that the flow of information could be improved, and he planned an 'Office of Address' on the model of an earlier French experiment under Cardinal Richelieu, which was to be a great clearing-house for scientific and technological information.[65] As so often in seventeenth-century England, however, such grandiose schemes came to nothing whereas Hartlib's less formal activity was crucial, and it was this that Oldenburg carried on: he even inherited a few of Hartlib's correspondents, notably the garrulous Somerset virtuoso, John Beale. Like Hartlib, Oldenburg assiduously communicated projects and discoveries from one scientist to another, providing an invaluable intelligence network without which, as contemporaries were only too aware, effort was easily duplicated or findings lost. Through his encouragement of minor virtuosi – often scattered in the provinces – Oldenburg was

[64] Hunter,[107] pp. 13–14.
[65] Brown,[101] pp. 31f; Webster,[72] pp. 67–77; H. M. Solomon, *Public welfare, science and propaganda in seventeenth-century France* (Princeton, 1972), chs. 2–3.

able to collect valuable information; he sometimes supplied them with instruments with which to make observations; and he helped teach them about the advancement of learning.[66]

Though Oldenburg had something in common with Hartlib in all this, he had greater influence due to writing his letters as Secretary of the Royal Society. The correspondence subsisted almost entirely by his efforts, but he wrote with the Society's authority, reporting opinions expressed at meetings on matters reported through his letters, and issuing instructions or suggestions in the Society's name about information to be collected or experiments tried. The rise in bulk of the Society's correspondence and the decline of corporate experiment were connected, because discussion of correspondence occupied an increasing amount of the Society's time. The Society's arbitration was critical, as is symbolised by the provincial inventor, Robert Lowman's, view of the Society as 'the Great Touchstone, and Judges of the probability and feasableness of experimentall designs'.[67] The Secretary depended on liaison with a London nucleus of high intellectual calibre, regularly including men as eminent as Hooke, whose opinions could be relayed to scattered enthusiasts at home and abroad.

Through the Society Oldenburg put pressure on scientists to divulge their findings for general consumption. In 1672 Newton disclosed his design for a reflecting telescope to the Society, commenting: 'since the Royal Society is pleased to think it worth the patronizing, I must acknowledg it deserves much more of them for that, then of mee, who, had not the communication of it been desired, might have let it still remained in private as it hath already done some yeares'. Oldenburg also tried to stimulate scientists to work on specific problems by reporting the progress of others: he sometimes even caused trouble by this, however salutary the general results.[68] The Royal Society also deserves credit for paying for the postage of these letters (or, in the case of some foreign ones, for getting them sent by official channels through the influence of well-placed Fellows). The expense of this was quite substantial, more than a private individual could afford, as Hartlib had found when his state pension dried up at the Restoration and he had had to suspend his activities.[69]

[66] For instruments, e.g. Childrey to Oldenburg, 22 Mar. 1669, Oldenburg,[121] v, 454; Oldenburg to Childrey, 6 April 1669, ibid, v, 477.
[67] Frank,[106] p. 95; Cl.P. VII (1), fol. 33.
[68] Newton to Oldenburg, 6 Jan. 1672, Newton,[127] I, 79; Hall,[113] p. 287.
[69] Hunter,[107] pp. 18, 21; Hartlib to Evelyn, 26 Nov. 1660, Evelyn, *Corr.*, no. 868.

So successful was Oldenburg in promoting correspondence that by the time he died he had established a tradition which his successors as officers were obliged to continue. Grew and Hooke undertook these epistolary duties in the immediate aftermath of Oldenburg's death in 1677, and others continued them later, perhaps particularly Hans Sloane, whose assiduity approached Oldenburg's though his letters still languish unpublished, mainly in the British Library. Such correspondence continued to exert a positive and salutary function, not least through the institutional backing the Society gave its officers. Like him, Oldenburg's successors collected information, publicised discoveries and acted as arbiters of the notions of scientific enthusiasts at home and abroad. All this was an achievement for which the Royal Society itself rightly took the credit.

Equally important was another initiative of Oldenburg's, which grew out of and even transcended his epistolary efforts, his inauguration in 1665 of the *Philosophical Transactions*. This owed something to the *Journal des Sçavans* that Denis de Sallo had begun in Paris earlier that year: but it differed from de Sallo's venture in devoting more space to science and less to book-reviewing, and it soon established itself as the leading European scientific periodical.[70] Initially Oldenburg founded the *Transactions* as a private venture and disclaimed the Royal Society's responsibility for the opinions expressed in its pages; he also ran the journal as a private speculation, supplementing his meagre official salary with the profits it made. But, though its unofficial status is sometimes pedantically pointed out, it was blessed with the Royal Society's general approval and was immediately associated with the Society by those who read it, becoming a public symbol of the Society's vitality.[71]

In some ways its influence surpassed that of the Society. It certainly reached a wider audience than did direct knowledge of the Society's activities: virtuosi, particularly in the country, often wrote to the publisher of the *Philosophical Transactions* rather than to the Society.[72] It was also badly missed whenever its publication was interrupted, as after Oldenburg's death or between 1687 and 1691.

---

[70] Brown,[101] ch. 9; for background see D. A. Kronick, *A history of scientific and technical periodicals 1665–1790*, second edition (Metuchen, N.J., 1976).
[71] Andrade,[116] pp. 13–14, 19; see also *Phil. Trans.*, 4 (1668), 911; Oldenburg,[121] II, xxv, v, xxv.
[72] E.g. Westmacott to the publisher of *Phil. Trans.*, 2 Sept. 1676, EL W.3, fol. 57;

John Beale thought 'the Royal Society does apparantly goe backwards till you have got an Industrious & ingenious Person to go on constantly with Phil: Trans:'.[73] The journal's indispensability was shown when Martin Lister continued to publish his work there even after breaking with the Society's meetings and ceasing to employ its imprimatur. Non-Fellows likewise used the *Transactions* to publicise scientific information, and the significance of the publication as against the Society's London activity is underlined by the case of Francis North, Baron Guildford. Guildford, an influential lawyer and a fairly active scientific enthusiast, used the *Transactions* but never sought election, for 'he could not discover what advantage of knowledge could come to him that way which he could not arrive at otherwise'.[74]

Much of the periodical's success stemmed from scientists' instant recognition of its value (in John Wallis's words) as 'a proper place for communicating new discoveries', or, as the preface explained in more detail when the series was revived in 1683 after a lapse, as 'a convenient *Register*, for the Bringing in, and Preserving many *Experiments*, which, not enough for a Book, would else be lost; and [they] have proved a very good Ferment for the setting Men of Uncommon Thoughts in all parts a work'.[75] Its influence as a 'Ferment' may be illustrated by an article by John Ray and Francis Willughby on the movement of sap in trees published in 1669, itself stimulated by earlier '*Queries*' in the *Transactions*. This has been described as 'the first systematic attempt to study the physiology of the living plant'; it opened up new fields of research, some of which were explored in subsequent issues and it is even possible that it inspired Grew to publish the work on plant anatomy that he had long been engaged on.[76] The circulation and influence of the *Transactions* was not limited to England: foreign scientists were pleased to see papers which they had sent to Oldenburg given wider currency in its pages, while Antoni van Leeuwenhoek found it a convenient place to publish his discoveries in microscopy.[77]

Smith to ditto, 5 Sept. 1676, EL S.1, fol. 105; Browne to ditto, 15 Dec. 1685, EL B.2, fol. 37.

[73] Beale to Evelyn, 6 April 1682, Evelyn, *Corr.*, no. 150; Weld,[94] 1, 322, 324–5.

[74] Carr,[142] p. 41; *Phil. Trans.*, 10 (1675), 310–11; Roger North, *The lives of the Norths*, ed. A. Jessopp (London, 1890), 1, 373–4, 383–4.

[75] Wallis to Oldenburg, 29 Mar. 1677, EL W.2, fol. 37; *Phil. Trans.*, 13 (1683), 2.

[76] Raven,[135] pp. 187–8; *Phil. Trans.*, 3 (1668), 797–9.

[77] Andrade,[116] pp. 20, 24–5.

Though many articles in the *Philosophical Transactions* were meant for practitioners in neighbouring fields (as is shown by certain abstruse mathematical treatises in Latin), the journal was also aimed at a wider public. As Oldenburg explained to the Belgian scientist René Sluse, 'these philosophical commonplace books are published in the English tongue because they are intended to be for the benefit of such Englishmen as are drawn to curious things, yet perhaps do not know Latin'. Hence in general the needs of English readers were given precedence over those of foreigners who might be 'clumsy in the English language' and therefore anxious for a Latin version.[78] A similar hint about audience was given when a recipe was inserted, 'several Curious Persons who either have not the leisure to read Voluminous Authors, or are not readily skilled in that Learned Tongue wherein the said Book is written, being very desirous to have it transferred hither'.[79]

Lesser men made clear how valuable they found the *Transactions* in keeping them up to date with scientific developments at home and abroad. The Durham virtuoso Peter Nelson waxed eloquent about the 'generall satisfaction' to which the periodical gave rise, while also requesting further details of recent views on important scientific questions.[80] Some contributions to the *Transactions* were significant stimuli to the investigations of minor men, such as Boyle's 'General Heads for a *Natural History of a Countrey*, Great or small', which were deliberately publicised thus, 'most men not knowing, what to inquire after, and how'.[81] Just how the periodical purveyed a scientific outlook is shown by a letter that Oldenburg published in 1676 from a Cambridge enthusiast whose identity has not been established. Though this was clearly his first contact with the Society, his interests and preoccupations had been formed by reading the *Transactions* and books recommended there. Conversely, such men could assist intellectual enterprise through its pages: Oldenburg saw its aim as not merely to record information '*but* also, and chiefly, to sollicite in all parts mutual Ayds and Collegiate endeavours for the farther advancement thereof'.[82]

[78] Oldenburg to Sluse, 2 April 1669, Oldenburg,[121] v, 469–70; Hoboken to Oldenburg, 29 July 1672, ibid, IX, 197.
[79] *Phil. Trans.*, I (1666), 125.
[80] Nelson to Oldenburg, 22 Aug. 1668, Oldenburg,[121] v, 23–4.
[81] *Phil. Trans.*, I (1666), 186–9; Oldenburg to Boyle, 27 Jan. 1666, Oldenburg,[121] III, 33.
[82] *Phil. Trans.*, 10 (1675–6), 533–41; ibid, I (1666), 163.

In fact, so far as it is possible to speak of the Royal Society's scientific standpoint, this is Oldenburg's point of view in the *Philosophical Transactions*, a more public version of that promulgated in his letters. Oldenburg consistently and fully expounded scientific aims and methods and, since these tracts bore the Society's name at their head and Oldenburg wrote letters on the Society's behalf, they naturally seemed to reflect its orthodox opinions. Oldenburg espoused a mature, Baconian concept of the proper role of science, placing a high value on the accumulation of natural histories but not despising the role of hypothesis, championing a moderate mechanistic view of the workings of the world and laying special stress on the achievements of Robert Boyle, 'that Noble Benefactor to Experimental Philosophy'.[83] By the time of Oldenburg's death this position was well established, and subsequent editors of the *Transactions* purveyed a moderate, Baconian line like Oldenburg's, though the focus of interest naturally shifted slightly with each editorial change. The journal continued, however, to popularise and stimulate scientific research.

The Royal Society was not the only agency articulating the science of the age. There were many less formal contacts between England and the continent and London and the provinces, and Oldenburg's was not the only correspondence spreading scientific ideas. The scientist and courtier Sir Robert Moray wrote frequently to Christiaan Huygens in the 1660s, while John Ray, whose links with the Royal Society were not of vast importance, could boast a wide network of contacts. So could John Aubrey and the mathematician John Collins – called 'the very Mersennus and intelligencer of this age' – who was like a subsidiary Oldenburg, knocking scientists' heads together, keeping provincial men informed and lending them books. A country virtuoso like Richard Towneley had contacts with numerous scientists although not closely associated with the Royal Society.[84] It has even been claimed that the Society neglected provincial virtuosi and that in the 1670s and 1680s Robert Plot had more success in orchestrating scientific activity in the provinces than the Royal Society had formerly. Plot was certainly active in sending out circulars

[83] Wood,[28] ch. 3; Kargon,[50] p. 135; *Phil. Trans.*, 1 (1665), 129.
[84] Huygens, *Oeuvres complètes*, vols. III–VI (The Hague, 1890–5), passim; Raven,[135] esp. p. 64; Hunter,[143] esp. p. 136; Bernard to Collins, 3 April 1671, Rigaud,[123] I, 159, and passim; Webster,[158] pp. 65 n. 28, 68.

and questionnaires and entering into correspondence with local naturalists, and he inspired much co-operation.[85] Also, in the 1680s and 1690s, the *Philosophical Transactions* was followed by other, less high-powered popularising journals, including the *Weekly Memorials for the Ingenious* and the *Athenian Mercury,* which served part of the function that it had hitherto served exclusively.

But, though this shows that the channels pioneered by Oldenburg were not unique, the Royal Society's significance is not reduced. It is even increased, since these alternatives were mostly secondary to or parasitic on the Society. Collins was an active Fellow and made assiduous use of the contacts this gave him. Plot's enterprise began as a private one (until he was officially encouraged by the Oxford authorities), but he deliberately obtained the Royal Society's approval to persuade people to help him.[86] Moreover he gradually came closer to the Society, until by the mid-1680s he was Secretary and editor of the *Philosophical Transactions*. He thus implicitly recognised the Royal Society as the institutional embodiment of his enterprise, with a continuity which individuals lacked, and his function was similar to that of Oldenburg earlier.

Similarly the journals of the 1680s and 1690s were positively derivative of the *Transactions,* and their proliferation a tribute to the success of their predecessor and the needs it had revealed. The *Weekly Memorials* was partly published to fill a gap left by the suspension of the *Transactions* in the early 1680s, and in the 1690s a new journal which appeared called *The Works of the Learned* was intended as a substitute for the *Transactions,* which was thought about to be given up.[87] Though such journals were aimed at a more general, casual audience than the *Philosophical Transactions,* they show a clear appreciation of that publication's importance.

The way in which alternatives enhanced the Royal Society is shown by another development of the 1680s, the inauguration of Philosophical Societies in Oxford and Dublin; a similar institution was also founded in Boston, Massachusetts, though plans for others in Cambridge and Aberdeen proved abortive. These were consciously modelled on the Royal Society, attempting to set up a forum where major and minor scientists could join in an educative and instructive

[85] Turner;[157] Gunther,[129] passim; for Plot's printed queries see Bodleian library MS Ashmole 1820a, fols. 222–5. [86] See below, p. 142; Birch,[118] III, 144.
[87] *Weekly Memorials for the Ingenious,* 1 (1682–3), preface; Walter Graham, *The beginnings of English literary periodicals 1665–1715* (New York, 1926), p. 27.

programme of experiment, discussion and correspondence. The Royal Society's example was also followed in providing formal organisation and elected officers and in levying subscriptions to pay for equipment for public experiments.[88] These trials were made, it was specified, 'not that they are all thought to be new, but for our satisfaction in matters whereof we are not fully certified', and this function of educating lesser virtuosi – different from, but overlapping with, its original Baconian ambitions – was clearly one served by the Royal Society, where public experiment also helped to popularise science.[89]

These imitative societies were deeply respectful of their parent body, having (in their own words) 'cheifly subsisted & grown up by the kind encouragement & countenance' of 'the most worthy Society in the learned World'. The Oxford group claimed that 'If . . . we may qualify our selves for your friendship & correspondence; and may in time be lookt on as one of your nurseries, it will be as much as we can desire.'[90] But though they had nothing but praise for the Royal Society, they illustrate an ambiguity in its position, for the older body suffered from their competition, as it may also have done from the new journals. Information was often only reported to the Royal Society after first going to Oxford, and the Society declined while its rivals thrived. Hence its council tried to capitalise on their success by a decree encouraging members of the Oxford and Dublin societies to join the London one at a concessionary membership fee.[91]

Indeed the Royal Society's achievement in promoting science caused repeated problems for its London nucleus, since it was easy to benefit from enterprises originated by the Society without being a regular member. This was already apparent in the 1670s, when the Society was experiencing severe institutional problems while scientific activity – much of it stimulated by the Society – flourished. When the Society's internal fortunes reached a crisis in 1673–5 a series of papers read by Fellows to the Society was published, including

88 Gunther,[120] pp. 45–6; Hoppen,[117] pp. 84f, 210–11; Stearns,[115] pp. 155f; Garden to Middleton, 17 July 1685, LBO x, fols. 227–8.

89 Musgrave to Aston, 1 Mar. 1684, LBO ix, fols. 130–1. Cf. Aston to Lister, 8 Mar. 1683, Lister 35, fol. 85.

90 Ashe to Southwell, 12 Feb. 1695, EL A, fol. 42; Gould to Aston, 6 Mar. 1683, LBO viii, fol. 298; Musgrave to Aston, 27 Mar. 1684, LBO ix, fol. 150.

91 E.g. Musgrave to Aston, 3 Mar. 1683, 20 May 1685, 11 Oct. 1685, LBO viii, fols. 324–5, x, fols. 157–8, 224–5; Birch,[118] iv, 402.

William Petty's *Discourse . . . Concerning the Use of Duplicate Proportion* (1674), apparently to give wide circulation to evidence of vitality at the Society's meetings.[92] The tension between institutional weakness and public success also underlies some very interesting 'Proposalls for the Good of the Royal Society' made at this time by Robert Hooke.[93]

Like others in these years, Hooke wished to cut the dead wood out of the Society and limit recruitment to those with a genuine concern for science. But he also felt it essential that the Society's benefits should be exclusively limited to its Fellows, so that active membership would be obligatory for any with serious scientific interests: 'there must bee somewhat to bee had by those that meete & are regular members which others must want'. To achieve this he would even have overturned the Society's Baconian commitment to the free dissemination of knowledge, arguing that 'Nothing considerable in that kind can bee obtain without secrecy because els others not qualifyed as abovesaid will share of the benefit.' He even wanted to limit the circulation of the Society's 'Gazets' to Fellows and to allow no one else to see them for twelve months. Hooke thus had visions of a Society so successful as an agency of communication that no one would any longer need to be a member, despite the 'Allurements' he enumerated, including

Desirable Acquaintance, Delightfull Discourse, Pleasant Entertainment by Experiments, Instructive Observations by Tracts, Considerable Intelligence by Letters, new discoverys by Inventors, Solution of Doubts & Problems, An easy way to know what is already known, Liberty to Peruse the Repository, the Letters & Registers, The Library, The Modaeles & instruments, Liberty to be present at mechanick, Optick, Astronomick, chymick, Physicall & Anatomick tryall.

This is an impressive-sounding list, but it is belied by the Society's institutional shortcomings that have already been indicated. Notwithstanding Hooke, the meetings for which he made such claims were really of little scientific importance, especially in comparison with the Society's more public role. As John Beale put it, 'dispatch and full communications are the life of the Royal Society'.[94] Its real achievement was less in the practice of science than in its organisa-

[92] Sharp,[144] pp. 270, 277.
[93] Cl.P. xx, item 50, fols. 97, 93v, 95. On these papers, their setting, implications and date, see a forthcoming article by Michael Hunter and Paul Wood.
[94] Beale to Boyle, 25 Jan. 1667, Boyle,[71] VI, 424.

tion. But the paradox of the early Royal Society is that this was not separable from its London meetings. The Society could not have consisted merely of a correspondence secretary and the editor of the *Philosophical Transactions,* even if they were the chief agents of its success: for their authority derived from association with regular meetings of a chartered institution. The two parts of the Society's operations had to co-exist, and this explains the tension between achievement and failure that it experienced throughout the century.

3

The scientific community

This chapter is about the social dimension of Restoration science: the sort of people who tended to become interested in natural philosophy, the discreteness of major scientists as a group, their relations with a wider body of enthusiasts, and the social and geographical extent of this broader dimension to the scientific advances of the age. What made a scientist, and why England was so exceptionally fertile of major talent in these years, are questions which have been often considered and can only be partially resolved, so complex are the considerations involved. Lawrence Stone has linked the triumphs of the new science with the expansion of educational facilities earlier in the seventeenth century, while R. K. Merton invoked the entire value-orientation of early modern English society.[1] Such imponderables are tackled here only rather obliquely, by indicating how scientific interests could be developed: for this shows what it took to become a member of the scientific community and the relative ease with which this could be achieved from different social backgrounds.

The juxtaposition of the burgeoning of science with the curiosity about personal memorabilia responsible for a rash of autobiographies and John Aubrey's *Brief Lives* reveals an early common characteristic of those who went on to scientific eminence. Several recalled their youthful curiosity about technical processes, inventions and the like.[2] More important is how this blossomed and the sort of people most likely to get the opportunity to take this inquisitiveness further. How many village Newtons never had such chances

---

[1] Lawrence Stone, 'The educational revolution in England, 1560–1640', *Past and Present*, 28 (1964), 80; Merton.[168]

[2] Paul Delaney, *British autobiography in the seventeenth century* (London, 1969); Aubrey,[161] I, 35–6, 409–10; II, 140; Flamsteed,[162] pp. xxiv, 10; see also Manuel,[139] p. 38.

is unknown, but it is useful to consider the social origins of those who played a prominent role in the pursuit of natural knowledge. Even if this only illustrates the inequality of opportunity in seventeenth-century England it is important in indicating the background, and hence the values and assumptions, of those at the centre of the scientific community.

Exactly who, however, are to be taken as the core of 'serious scientists'? Any selection is bound to be somewhat arbitrary, raising problems of definition which will be examined later. But it is possible to pick out those whose contribution to science seems outstanding, and for this purpose we shall take those sixty-five British-born scientists active in late seventeenth-century England who were deemed worthy of inclusion in the recent *Dictionary of scientific biography*. The *Dictionary*'s editors are naturally open to criticism for a few of their inclusions and omissions but these do not invalidate the sample offered: since their criteria were intellectual not social, there is no reason to suspect a serious distortion in the percentages derived from different social classes. Moreover deductions based on the *Dictionary* can be checked by considering an independent analysis of the social origins of a slightly larger group of men selected for their scientific achievement which came to very similar conclusions.[3]

Of the men in the *Dictionary*, the greatest number – 40 per cent – came from landed families, including not only the aristocrats Boyle and Brouncker but others like Martin Lister and Edward Tyson. This is not entirely surprising, since more opportunities were open to this class than any other: it provided 33 per cent of all seventeenth-century entrants to Cambridge although comprising less than 2 per cent of the population.[4] The next largest group – 23 per cent – was of sons of the Anglican clergy; these included Walter Charleton, Robert Hooke and Christopher Wren. This class is again disproportionately common, even more so than with the aristocracy and gentry, for the clergy and their families made up under 1 per cent of the population. If a link is sought between the Reformation and the rise of modern science, it is perhaps to be found in clerical marriage rather than theology.

Other professions account for a further 5 per cent, while the sons of merchants made up 12 per cent of the sample, among them Isaac

---

[3] Hans,[320] table 5. The information provided in Gillespie[134] has been supplemented where necessary from standard reference books.
[4] Cressy,[131] p. 312 (based on a four-college sample).

Barrow and Henry Power. Artisans, yeomen and others of lesser social status contributed the remaining 14 per cent (other than 6 per cent unknown): William Petty, son of a Romsey clothworker, and John Ray, whose father was a blacksmith, fall into this category. The fact that men of such lowly backgrounds could advance themselves educationally to become the peers of others of more exalted origins indicates the relatively wide spread of educational opportunity in seventeenth-century England at all but the humblest level, which has often been commented on. But, even if open to many, education tended to be disproportionately the prerogative of a very small minority of the country's inhabitants: hence most scientists came from the landed and professional classes who made up less than 5 per cent of the population while only a handful came from the classes that formed its bulk.[5] This predominance and the relative fewness of merchants' and artisans' sons is also a reminder that if seventeenth-century English science is to be seen in any sense as 'middle class', it was not really 'bourgeois'.

The early development of scientific interests depended on a combination of favourable circumstances and self-help. The story of John Flamsteed, the future first Astronomer Royal, is indicative of the surprising ease with which science could be pursued in an unlikely environment. The son of a Derby maltster, Flamsteed learned the elements of arithmetic from his father and managed to find and read books by mathematicians of the early seventeenth century. He was encouraged by local gentlemen who lent him books and encouraged his precocious interest in astronomy, making up for his failure (to his great regret, due to ill-health and his father's hostility) to attend university. In 1669–70, when in his mid-twenties, Flamsteed established relations with Oldenburg and Collins, and the communications network of the Royal Society worked for him as for others: he made contact with London scientists and also, nearer home, with the circle of Richard Towneley in Lancashire, where Flamsteed was able to develop his interests in a sympathetic environment. As satisfactory biographies of other scientists are completed, they reveal parallel experiences: Christopher Wren grew up in a family with an interest in science, while William Petty rose from his depressed artisan origins to make his fortune and to gain an education abroad.[6]

[5] Stone, 'The educational revolution', esp. pp. 67–8; Cressy,[131] p. 312 and passim.
[6] Flamsteed,[162] pp. 9–12, 20–4, 28–30; Webster,[158] pp. 68–70; Bennett,[141] pp. 30f; Sharp,[144] ch. 1.

Flamsteed was unusual among major scientists in his lack of a university training, and most nurtured (or first gained) their interest in science at university. This is again illustrated by the statistics from the *Dictionary of scientific biography*: no less than 75 per cent of all those included were educated at Oxford or Cambridge and 6 per cent more gained an academic education elsewhere. Since the *Dictionary* is most susceptible to criticism for including some dilettantes of literary rather than scientific merit (such as Izaak Walton), this overwhelming majority is all the more striking. Hardly less important is the fact, illustrated by another recent study, that among scientists men who actually finished their degree predominate, whereas most of those who attended university in the seventeenth century never bothered to take a degree.[7]

The universities' encouragement of science, which has often been underrated, will be examined in chapter 6. The undergraduate course was less important in this connection than for instilling techniques and mental attitudes supplementing those (such as Latinity) which most had already acquired at school; contacts were also encouraged between like-minded individuals which they carried on to later life. Scientific study in its own right was more frequently pursued at a graduate than an undergraduate level: here budding scientists obtained a training in research and an acquaintance with the new philosophy, while discovering the value of learning for its own sake.[8] The men in the *Dictionary of scientific biography* displayed a quite disproportionate propensity to take further degrees in addition to first ones: 55 per cent fall into this category, though this includes a number of degrees awarded in later life. It was thus through further and higher education that people tended to become 'scientists', and this created a kind of community, giving a shared background in academic life which scientists carried over to their subsequent occupations. Graduate study could also lead directly to a profession closely related to science, for many qualified for medical degrees, either at Oxford or Cambridge or at European universities like Padua and Leiden.

Hence the occupational structure of the men who stand out as scientific innovators displays a strong medical and academic streak. Doctors form the most numerous group (35 per cent), including men like Nehemiah Grew, Martin Lister and Hans Sloane. Professors and

[7] Frank,[132] p. 198. See also Hans,[320] table 4.
[8] Frank,[132] pp. 197–9, 261–3 and passim; on schooling see Vincent,[214] esp. ch. 4.

professional scientists are next most common (23 per cent): Robert Morison, Isaac Newton and John Wallis held chairs in scientific subjects at Oxford and Cambridge, while the new scientific institutions provided employment for Flamsteed and Hooke. If this category is extended to include those who had been academics in the Interregnum but took preferment in government or the church after 1660 – such as Seth Ward and Christopher Wren – it swells almost to equal the doctors (32 per cent). There can be no doubt that by the late seventeenth century science was becoming a professional activity in its higher reaches, in the sense that it was pursued largely by men with careers in closely related fields.

Those whose contribution to science was outstanding do not all, however, fall into this class. What made such men scientists was not their academic or medical status but the attitudes that they had acquired in preparing themselves for it, and they shared these with others with a similar education but different vocations. The roll call of scientists includes the aristocrat Boyle, the gentleman Richard Towneley, the courtier Lord Brouncker and the clergyman Robert Sharrock.[9] Such men found time for science either in leisure derived from private means or in spare time from careers in the service of God or the King. But they shared with their colleagues in scholarly and medical callings the same academic skills and the same familiarity with up-to-date research on the subjects that interested them and to which they contributed. Participation was in principle open to anyone who acquired the relevant techniques: in other words, to any educated person.

The exact boundaries of the class are therefore very hard to draw, and this illustrates the problem of defining 'scientists' in the seventeenth century so far avoided by reliance on the sample from the *Dictionary of scientific biography*. There is obviously no doubt that Newton, Boyle or Hooke were 'scientists'. But as one moves away from such intellectual giants the problem of definition becomes greater and a subjective element enters, depending largely on varying assessments of individuals' scientific achievement. Thus opinions have differed as to whether William Petty should be classed as a 'serious scientist' and it is doubtful if a rigid categorisation is helpful, for Petty was undeniably deeply involved in the new philosophy, whatever his contribution to theoretical advance in purist

[9] Such men make up 42 per cent of the total if the former academics just mentioned are included, 32 per cent if they are not.

disciplines.[10] What is important is the essential bond between all intellectuals with a commitment to the new philosophy, whether original or derivative, professional or amateur.

In some scientific subjects more precision is possible. The higher echelons of mathematics – then as now – excluded all but a handful. The chief importer of continental books found 'that England doth not vent above twenty or thirty of any new mathematical book he brings over', and it was agreed that Newton's *Principia* was not widely understood. Even a scientist like William Molyneux of the Dublin Philosophical Society found it heavy going, and all or part of it was immediately 'translated' into plain English for interested non-experts from the King downwards, by John Locke and others.[11] In anatomy and physiology a high degree of expertise was similarly essential. Other topics, however, required less skill, like natural history and the earth sciences: this is illustrated by analysis of discussions of such matters at the Royal Society. Here there was much more overlap between scientific innovators and a larger body of enthusiasts who shared a similar – if less advanced – educational background, taking a close interest in scientific discovery even if contributing only marginally themselves. These were the so-called 'virtuosi', and it is significant that though they are sometimes differentiated from the elite of scientists, contemporaries used the term 'virtuoso' to describe both major and minor men with shared interests.[12]

The strong Baconian impulse in Restoration science itself tended to open up the scientific community to a large body of minor assistants in addition to the few who had mastered highly technical disciplines. Sprat stressed the value of the contributions of lesser men in his *History of the Royal Society*, arguing that 'though many of them have not a sufficient confirmation, to raise *Theories*, or *Histories* on their *Infallibility*: yet they bring with them a good assurance of likelihood, by the integrity of the *Relators*; and withall they furnish a judicious *Reader*, with admirable hints to direct his Observations'.[13] Though the significance of more high-powered contributions should not be underrated, it is easy to forget the genuine

[10] On this matter, see Sharp,[144] pp. 157–8; Rattansi,[103] pp. 129–31.

[11] Collins to Beale, 20 Aug. 1672, Rigaud,[123] 1, 200; Hoppen,[117] pp. 124–5; Douglas McKie, 'James, Duke of York, F.R.S.', *N.&R.R.S.*, 13 (1958), 7–8; *Phil. Trans.*, 16 (1687), 291–7; Axtell,[133] pp. 166–72, 175–8.

[12] Frank,[106] p. 93; Westfall,[295] pp. 13–14.   [13] Sprat,[92] p. 195.

levelling effect of Baconianism. The demands of induction gave minor observers a role they had been denied when deduction *a priori* dominated intellectual life, as under the old scholastic regime from which the new scientists were so eager to extricate themselves, and it conferred significance on lesser thinkers whom one might otherwise dismiss.

The unity of this wider scientific community, comprising lesser as well as major men, is illustrated by contemporary study of the movement of tides: this was linked to the widespread interest in gravitation underlying Newton's work on the *Principia*. The immediate background to Newton's theories was informed discussion among men familiar with the most technical scientific traditions, like Wren and Hooke. But a hypothesis by John Wallis about the relationship of the ebb and flow of the sea to the common centre of gravity of the earth and the moon revealed great related curiosity at a less high-powered level which contributed to the eventual outcome of the debate. Both when Wallis's paper was read at a meeting of the Royal Society and after its subsequent publication in the *Philosophical Transactions* in 1666, this topic aroused widespread interest among virtuosi, and when the ensuing discussion made Wallis appeal for empirical observations of tidal variations in different places, such men provided them.[14] As usual the Royal Society acted as orchestrator, publishing inquiries in the *Transactions*, and observations collected or supplied by virtuosi about the tides at Bristol, Chepstow and Plymouth afforded the data which Newton used in his discussion of their mechanism in the *Principia*.[15]

Lesser men were chiefly important for collecting information, but they were not hesitant in suggesting rival theories concerning this and other natural phenomena, theories which major practitioners had to take seriously. Wallis became engaged in a lengthy correspondence on the subject with a country virtuoso, Francis Jessop of Sheffield, and even if Jessop's theory was eccentric, others (such as the Dorset parson Joshua Childrey) produced really cogent objec-

---

[14] Bennett[48]; *Phil. Trans.*, 1 (1666), 264–89; Powle to Oldenburg, Sept. 1666, Oldenburg,[121] III, 235–6.

[15] *Phil. Trans.*, 1 (1666), 311–13; Deacon,[182] pp. 98f; Birch,[118] II, 111, 119, 121–2, 127, 134–5, 137; Oldenburg to Powle, 12 Sept. 1666, Oldenburg,[121] III, 228–9; Newton,[63] pp. 438–9, 479–81, 584–7, 588–9.

tions to Wallis' ideas.[16] As well-informed commentators on the work
of scientists the virtuosi are easily underrated. This was the function
of men like Sir John Hoskyns, a close friend of Hooke and a very
active Fellow (and sometime President and Secretary) of the Royal
Society. References in Hooke's *Diary* and the Society's minutes
reveal Hoskyns' fertility in constructive suggestions, though he hardly
carried out any scientific work himself (his main output was of sets
of 'Enquiryes'). Occasionally he came up with really ingenious ideas
that might have influenced scientific work, while Hooke thought his
notion for linguistic reform 'very good most simple'.[17] He can be
paralleled by the schoolmaster Ralph Johnson, a zoological expert
respected by Ray, who claimed him as the source of the fruitful
suggestion that his catalogue of British plants be arranged in natural
rather than alphabetical order.[18]

The overlap between 'real' and 'amateur' science is further illustrated by the links between the Royal Society's 'repository' and the
virtuoso 'cabinets' of which London was full. These are often written
off as miscellaneous and trivial, and in retrospect this seems an
accurate judgement on many of their oddities uncritically treasured
in a manner typical of the dilettante virtuoso movement. Such is the
predominant impression left by the surviving records of the objects
bought and sold by the broker and collector, William Courten, and
by reports of the fossils and exotic creatures in the collection of one
Captain Hicks, who made his living by furnishing ladies' closets with
different kinds of shells.[19] But serious scientists like Ray and Lister
nevertheless expressed great interest in these museums and the objects they contained (Courten's collection was later bought by Sloane
for his) while the Royal Society paid £100 for one of the most famous
such cabinets, that of Robert Hubert, of which a catalogue had been
published.[20] Hubert's rarities formed the basis of the Society's repository, which sounds miscellaneous enough from the highlights in
it itemised by the Italian visitor, Lorenzo Magalotti, including 'an

[16] See Jessop's letters of 1673–5 in Oldenburg,[121] x, passim, and his papers in EL I.1,
      fols. 165–6; Deacon,[182] pp. 102f; *Phil. Trans.*, 5 (1670), 2061–8.
[17] Cl. P. XIX, fols. 1, 3, 32; Hooke,[165] p. 393 and passim; 'Espinasse,[136] pp. 108, 126;
      Hunter,[143] ch. 2; Birch,[118] I, 155–6 and passim; Aubrey 12, fols. 187–231.
[18] Raven,[135] p. 249. See also ibid. p. 319.
[19] Sloane 3961–2, 3988; Lodge to Lister, 6 Feb. 1674, n.d., Lister 34, fol. 151, Lister
      3, fol. 169. For a general account of such cabinets see Candill,[147] chs. 4–5.
[20] Raven,[135] p. 229; Ray to Lhwyd, 7 May 1690, Ray,[125] p. 207; Lodge to Lister, n.d.,
      Lister 3, fol. 169; Birch,[118] II, 64; Rober Hubert [alias Forges], *A Catalogue of the
      Many Natural Rarities* (London, 1664).

ostrich, whose young were always born alive; a herb which grew in the stomach of a thrush; and the skin of a moor, tanned, with the beard and hair white'. Yet in 1681 a lavish and scholarly catalogue of the extended collection was published by Nehemiah Grew on the Society's behalf, a clear expression of the scientific value attached to it. Such a collection made it easy 'to find likenesse and unlikenesse of things upon a suddaine' (in Sir John Hoskyns's words), while Oldenburg stressed how even private museums 'will at length make up such a Store-house, as our Society designeth for an Universal History of Nature'.[21]

There were some contrasts between the Society's cabinet and its virtuoso forebears. Grew's catalogue is a reminder that an institutional collection could be permanent, unlike private ones, and those solicited to present objects to the Royal Society were urged that their gifts would be carefully preserved for posterity 'probably much better and safer, than in their own private Cabinets'. This was equally true of the Ashmolean Museum at Oxford, which was based on a similar collection, and John Aubrey retrieved objects that he had deposited with friends to assure their survival there.[22] Approaches to such facilities also differed, as is shown by Robert Hooke's view that 'the use of such a Collection is not for Divertisement, and Wonder, and Gazing, as 'tis for the most part thought and esteemed, and like Pictures for Children to admire and be pleased with, but for the most serious and diligent study of the most able Proficient in Natural Philosophy'.[23]

Indeed the values of at least some virtuosi were antipathetic to the pursuit of serious science. Bacon had criticised their tendency to trivial curiosity and their stress on the social esteem of knowledge, though paradoxically his call for collaboration frequently authorised what he disapproved.[24] This unconstructive mentality is illustrated by many of John Evelyn's comments in his *Diary* on proceedings at the Royal Society: he often noted curiosities that struck him as 'rare' and 'wonderful', and tended to ignore more serious aspects of the Society's business; Pepys was similarly preoccupied with gadgets and the like. In fact there was a certain tension even within

---

[21] Magalotti, *Travels of Cosmo III, Grand Duke of Tuscany, through England (1669)* (London, 1821), p. 188; Grew, *Musæum Regalis Societatis* (London, 1681); Hoskyns to Aubrey, 25 Mar. 1674, Aubrey 12, fol. 214; Oldenburg to Lister, 21 Oct. 1671, Oldenburg,[121] VIII, 307.

[22] *Phil. Trans.*, 1 (1666), 321; Hunter,[143] p. 43 n. 5 (cf. Aubrey 2, fol. 26; Aubrey,[161] II, 166).   [23] Hooke,[65] p. 338.   [24] Houghton,[146] pp. 55n, 56, 72.

the virtuoso movement between a Baconian impulse to instructive-
ness and utility and a proneness to inconclusive and frivolous curiosity which writers on the subject deprecated, stressing the need for
judgement and practical knowledge.[25] Lesser men were frequently
glad to collect information but reluctant to systematise it. 'Meere
compiling will content mee', was Hoskyns's statement of this attitude, so that their contribution to the advancement of learning often
had severe limitations.[26]

But serious scientists could not escape their association with the
virtuosi even if they wanted to (and there is little evidence that they
did). Besides their role as Baconian collectors and arbiters, the
virtuosi provided the staple attendance and finance on which a
formal body like the Royal Society depended. The Society's organisers showed their awareness of this when trying to rationalise the
membership by expulsions in the 1670s and 1680s, arguing the need
to retain a membership larger than that of active 'scientists'. The
virtuosi also made themselves useful to the Society by serving as
officers.[27] Minor claims can thus be made for the virtuosi as patrons
of scientific investigation, while they also provided a significant
market for scientific books and equipment.

As for scientific instruments, and not least for those newly introduced in the seventeenth century like the telescope, the microscope,
the barometer and the thermometer, the demand from the virtuosi
was sizeable and not uncritical. Typical in his high standards was Sir
George Croke, High Sheriff of Oxfordshire, who insisted on
employing 'the best Workman' to get a telescope that was 'one of the
most Exact for its length in England', and many virtuosi could boast
outstanding equipment for which they were prepared to pay
handsomely.[28] A handful of scientists could never have provided
enough custom to support manufacturers whose expertise soon
outstripped the abilities either of ordinary lens-makers and glass-
blowers or of scientists who prepared their own equipment. The
larger 'virtuoso' market, on the other hand, enabled a few craftsmen

[25] Evelyn,[164] III–IV, passim; Nicolson,[148] ch. 1; Caudill,[147] esp. ch. 1, though he fails
adequately to bring out this tension, which I shall explore in my forthcoming study
of John Evelyn.
[26] Hoskyns to Aubrey, 16 Mar. 1678, Aubrey 12, fol. 220.
[27] Hunter,[107] pp. 16–21 and passim.
[28] Croke to Oldenburg, 23 Feb. 1674, Oldenburg,[121] X, 484–5; cf. Croke to Oldenburg, 2 Feb. 1674, ibid, X, 461, and e.g., Robinson to Power, 4 Mar. 1660, Sloane
1326, fol. 101v.

to specialise and so refine the devices that they supplied to a few innovating scientists at home and abroad: the names of workmen like Richard Reeves, Christopher Cocks, John Yarwell and John Melling, familiar from the tributes of active researchers, also recur in the correspondence of the virtuosi.[29]

Equally important, virtuosi could contribute to the costs of producing books. Publication by subscription was beginning to become normal at this time, looking forward to the immense extension of the practice in the eighteenth century. It was seen as a way to get into circulation worthwhile books which it was widely held the Stationers' Company (which held the monopoly of printing) was too mercenary to publish, and the role of subscribers was really valuable, as contemporaries saw. Numerous enterprises were similarly organised in seventeenth-century England, and it is not entirely fanciful to compare the virtuosi with the investors in Joint Stock companies who played a critical role in economic life.[30] Complaints about the Stationers were widespread: 'the Booksellers at London are wholly bent upon present gain', Martin Lister felt, 'I confess that the greatest part of Natural Historie has been starved and abused by the Avarice of Stationers who have beat down the Artist'.[31] Subscribers offered an alternative, particularly with expensive illustrated books which were vital to science. Works published thus included Grew's *Musæum Regalis Societatis* (1681) and his *Anatomy of Plants* (1682), the first volume of Ray's *Historia Plantarum* (1686) and Edward Lhwyd's *Lithophylacii Britannici Ichnographia* (1699). When Willughby's *Historia Piscium* came out with the Royal Society's encouragement and with plates paid for by various virtuosi in 1686, Ray confessed that 'I did before despair of any Booksellers medling with it unless encouraged by subscriptions.'[32]

To analyse the composition of the whole scientific community, therefore, one has to move away from the elite of scientists with European reputations and predominantly professional status to a

---

[29] I have developed this theme at greater length in an unpublished paper given at a symposium on scientific instruments held by the British Society for the History of Science and the National Maritime Museum at Greenwich in August 1977.

[30] S. L. C. Clapp, 'The beginnings of subscription publication in the seventeenth century', *Modern philology*, 29 (1931), 199–224; F. J. G. Robinson and P. J. Wallis, *Book subscription lists: a revised guide* (Newcastle upon Tyne, 1975).

[31] Lister to Aston, 1 Feb. 1682, LBO VIII, fol. 261.

[32] Robinson and Wallis, *Book subscription lists*, pp. 1–2; Bluhm,[108] p. 102; Ray to Robinson, 13 Mar. 1685, Ray,[125] p. 142; see above, p. 42.

larger and rather more inchoate group of 'virtuosi'. These lesser collectors, theorists and enthusiasts were familiar to contemporaries, as is shown by satirical comment about them, notably in Thomas Shadwell's play *The Virtuoso* (1676) and in verses by Samuel Butler which must be typical of a wider genre. Much of this is lost, but it even embarrassed Hooke, and it is symptomatic of the absence of precise boundaries between 'Scientists' and 'virtuosi' that major and minor men were pilloried together.[33] The very size and diffuseness of this broader group, however, mean that its social breakdown has to be much more impressionistic. Some useful sources of information on the subject have yet to be properly explored, such as the published lists of subscribers to scientific books. These, if analysed, would afford an interesting view of those who patronised natural philosophy throughout the country, even if the purchase of expensive volumes is not always evidence of serious concern for science.

The one group of virtuosi who have been assiduously studied are the Fellows of the Royal Society, for the printed membership lists of that body give a fixed sample convenient for statistical analysis which has attracted a number of prosopographical studies. Some scholars have assumed that election to the Society meant more than it often actually did: it has long been known that the Society's connections with the establishment gave it a social *éclat* which attracted many to its ranks for frivolous reasons, and membership is not necessarily proof of an active commitment to science. These difficulties can be partly overcome by considering not just all who joined the Society but those active in it, however, and such qualifications do not destroy the interest of the social sample represented by the Society's Fellows.[34]

The Society's early membership was dominated by the professional and landed classes, and (particularly in its earliest years) it was closely connected with the court. Though meetings were held at Gresham College in the City of London for reasons indicated in the last chapter, the Society's centre of gravity was much more Westminster and Whitehall than the City – the court, the Houses of

---

[33] Nicolson,[148] ch. 3 and appendix; Hooke,[165] p. 235; Lloyd.[287] The case argued in J. M. Gilde, 'Shadwell and the Royal Society: satire in *The Virtuoso*', *Studies in English Literature*, 10 (1970), 469–90, which differs from the view expressed in the text, seems to me highly unconvincing, as consideration of Lloyd's evidence will reveal.   [34] Hunter,[107] esp. pp. 32–42.

Parliament and Westminster Hall, the seat of justice, rather than the City's commercial exchanges. There were practically no tradesmen in its ranks and surprisingly few merchants in view of the hopes of Thomas Sprat and others that the new science would be patronised by active 'men of *Trafick*'.[35] Sprat's aspirations have too often been uncritically echoed by modern scholars, who have thereupon conjured up misleading visions of science and commerce advancing hand in hand. In fact merchants and tradesmen together comprised only 6 per cent of the Fellows in the period 1660–85, and such merchants as belonged were exceptional in their class, having courtly connections, lodgings at Gresham College or the like. The staple of the Society's membership were doctors and professional scholars (31 per cent), aristocrats and landed gentlemen (30 per cent), and a variety of men associated with the court – politicians, courtiers, diplomats, holders of more or less sinecure offices (16 per cent), and, in their train, an admixture of government officials (5 per cent). There was also a steady mingling (7 per cent) of churchmen.

On the face of it, therefore, the membership was predominantly recruited from the professions, land and government, and was not particularly mercantile. In this it was typical of the more general leisured culture of London, which filled the coffee-houses and theatres with cultivated, well-informed dilettantes with a wide range of interests. There are hints that active Fellows of the Society often engaged in entrepreneurial activity as a side-line to their main career. Sir John Lowther's estates at Whitehaven in Cumberland provided more than half the coal for Dublin; Thomas Povey was a government official with business interests; the doctor Daniel Coxe had extensive colonial commitments (he was singled out in this connection by Daniel Defoe in his *Essay upon Projects*); the lawyer Sir John Clayton was involved in a pioneering scheme for constructing lighthouses; even Martin Lister had 'great hopes of considerable mines in my own lordship'.[36] Too much should not, however, be made of this, since such concerns are not always well-evidenced and are difficult to summarise statistically: it is impossible to know whether the patrons of drama showed similar tendencies. The fact that trade was at most a subsidiary interest shows where the true centre of gravity of

---

[35] Sprat,[92] pp. 129–30.
[36] Petty to Southwell, 28 April 1677, Lansdowne,[235] p. 26; Jacobsen,[225] pp. 43–9; Defoe,[202] p. 29; E. and D. S. Berkeley, *John Clayton* (Chapel Hill, 1963), p. 12; Lister to Ray, 22 Dec. 1669, Ray,[124] 50; Carr,[142] p. 273.

virtuoso culture lay, suggesting a mentality like that of the English aristocracy before the Civil War, with its 'entrepreneurial' characteristics but very mixed values. At most it illustrates the significant fact that in English society there was no real hostility to trade and industry properly pursued: if it offers any clues to the motivation underlying enthusiasm for science, it is the background to the 'indolently utilitarian' attitudes to be surveyed in the next chapter. The exception who proves this is the financier, Sir John Banks, whose election to the Royal Society has been linked to his aspiration to escape his commercial background for the elevated company of governmentmen and savants who swelled the Society's ranks.[37]

If this establishes where the Society found the financial and moral backing to perform its functions in the world of science, it is less clear what it proves about support for the new science as a whole. Only a minority of scientific enthusiasts belonged to the Society and not all the reasons why others failed to join were random. The Society's bias towards London rather than English science as a whole was noted in the last chapter. But membership was far from inclusive even of those with known scientific interests in the metropolis, and it is not difficult to find virtuosi who pursued natural philosophy without belonging to the Society.[38]

The reasons for failing to seek election were often fairly random, but in other cases they were not. Thus no woman ever belonged although there is considerable evidence of feminine curiosity about science, from the notorious Margaret, Duchess of Newcastle – authoress of various idiosyncratic scientific works – downwards: one academy proposal even suggested of its female pupils that 'those that think one Language enough for a Woman, may forebear the Languages, and learn onely *Experimental Philosophy*'. The Duchess was lavishly entertained by the Society in 1667, but it was never suggested that she be elected.[39] Equally significant is the evidence of the Society's social exclusiveness. The government clerk, John Collins, claimed that he was elected 'though a mean person, for his knowledge in Mathematics': he thus implied that others of similar status and lesser intellectual achievement were debarred from

---

[37] Lawrence Stone, *The crisis of the aristocracy* (Oxford, 1965), ch. 7 and passim; Coleman,[145] pp. 135–9, 196–7.

[38] Hunter,[107] pp. 12–13.

[39] Caudill,[147] p. 330; D. Grant, *Margaret the first* (London, 1957), esp. chs. 1, 10; Meyer,[150] ch. 1; Nicolson,[148] pp. 104–14.

membership altogether, if they were not deterred by the two guinea subscription from which Collins had to be exempted.[40]

We must therefore make an attempt to judge how far the Royal Society was representative of the general enthusiasm for science in its preponderance of professional and landed Fellows and its relative lack of merchants and artisans. Was this the result of self-conscious and artificial exclusiveness, or did science genuinely tend to interest some classes rather than others? It is essential to distinguish here between a curiosity about a whole range of scientific topics and a more single-minded technological concern with applied knowledge. As we shall see in the next chapter, the latter had an audience that went well down the social scale among men who bought technical handbooks on agriculture or shipbuilding or primers in elementary mathematics: on medicine, too, there seems to have been a wide audience for a vernacular literature, which did not necessarily imply broader intellectual curiosity.[41] On the whole, however, a devotion to more varied scientific activities does not seem to have spread so far and was primarily associated with leisure, except for those (such as doctors) whose careers brought them into contact with it. In other words, an interest in science was most often found in the professional classes, the gentry, and what Alan Everitt calls the 'pseudo-gentry', people of private means, usually in an urban environment, who lived the life of leisure traditionally associated with the landed classes although not landed themselves.[42]

Clearly this was the principal clientele of the coffee-houses and other venues in London where – in potential rivalry with the Royal Society – an interest in science could be cultivated. People of this background were probably also the patrons of the public lectures on scientific topics delivered in Restoration London: the fee of three guineas charged for a course on chemistry in the 1690s is indicative of the wealthy audience hoped for. It was the gentry who were blamed for neglecting the lectures given by eminent scholars to dwindling numbers at Gresham College (whether this was a new departure at the Restoration for that ailing institution is not

---

[40] *C.S.P. Domestic, 1677–8* (1911), p. 543. Cf. Hunter,[107] p. 12, F235.
[41] See below, pp. 109–10; Webster,[72] pp. 264–73.
[42] Everitt, 'Social mobility in early modern England', *Past and Present*, 33 (1966), 70–2, and 'Kentish family portrait: an aspect of the rise of the pseudo-gentry' in C. W. Chalklin and M. A. Havinden, eds., *Rural change and urban growth* (London, 1974), pp. 169–71.

certain).[43] In addition, contemporary accounts make it clear that the nobility and gentry were the main clients of the educational facilities available in London on scientific, technical and humane subjects and courtly accomplishments, which had earned the metropolis the title of 'Third University' earlier in the seventeenth century.[44]

Merchants showed some curiosity in acquiring exotic objects while abroad – there was a fine collection at the East India House in the City – and they sometimes gave or offered such items to the Royal Society.[45] But members of this class apparently lacked the leisure or inclination for sustained scientific pursuits. Attention has been drawn to the continual application required for success in commerce, which made it hard for active merchants to take a continuous interest even in politics: the difficulties of identifying a self-conscious 'bourgeois' class in early modern England are notorious.[46] The same was true of science, which was patronised only by those, like the citizen Abraham Hill, whose wealth was based on trade but who lived a life of leisure and hence could afford the time required. Active merchants not only had no opportunity for science but little interest in it. The benefaction to Hooke of the businessman, Sir John Cutler, proved problematic, and the Royal Society was quickly disillusioned in 1673 when it moved its meeting-place from Arundel House to Gresham College, encouraged not least by 'the hopes, which they find grounds to entertain, of meeting with some considerable benefactors at that end of the city'.[47]

Moving down the social scale to artisans and tradesmen, an interest in natural philosophy is found only among those whose business had some scientific connection: manufacturers of scientific instruments, almanac-makers and apothecaries, that lower house of London medical practice whose feud with the College of Physicians

43 Houghton,[201] 4 (1694), no. 89; Hans,[320] p. 153; E. G. Forbes, ed., *The Gresham lectures of John Flamsteed* (London, 1975), p. 2.

44 Richard Burton, *Historical Remarques, and Observations* (London, 1681), part 1, p. 109; Thomas De-Laune, *The Present State of London* (London, 1681), pp. 160–1; Caudill,[147] p. 356.

45 Magalotti, *Travels of Cosmo III*, pp. 325–7; Birch,[118] III, 397; Weld,[94] I, 349–50; Grew, *Musæum Regalis Societatis*, p. [388]; Brooke to Lister, 26 Sept. 1672, Lister 34, fol. 63v.

46 R. Grassby, 'English merchant capitalism in the late seventeenth century', *Past and Present*, 46 (1970), 87, 104; Jones,[222] pp. 14–15. See also J. H. Hexter, 'The myth of the middle class in Tudor England' in *Reappraisals in history* (London, 1961), pp. 93–9.

47 Hill,[122] p. v; 'Espinasse,[136] pp. 4, 83; Birch,[118] III, 100.

is so prominent in these years.[48] Among apothecaries, one might mention John Conyers – though he was unusual in being related to the physician Francis Glisson – who collected a museum of rarities that formed the basis of Sloane's and who wrote papers on scientific topics. There is also Hooke's acquaintance, Tom Hewk, and James Petiver, a collector who became a Fellow of the Royal Society in 1695.[49] Instrument-makers were curious about mechanical, magnetic and astronomical phenomena: virtuosi and others gathered at their shops to talk or to watch eclipses and Hooke's splinter group from the Royal Society met at the premises of William Bennet, a clock-maker. Hooke's *Diary* shows his frequent contacts with instrument-makers, at least some of whom (such as the famous clock-maker, Thomas Tompion) discussed a great variety of subjects.[50] The almanac-making class was wider, ranging from artisans to doctors and clergymen. But among those of lower rank one can single out Henry Coley, a tailor by trade, who made almanacs in which (like other similar writers) he tried to disseminate some elementary astronomical information to his readers – with what success is not known. His surviving notebooks show his desire to understand the intellectual basis of the predictions he purveyed, and he also taught mathematics.[51]

But these men often had rather narrow attitudes, evidently feeling themselves incompetent to consider topics beyond their immediate speciality, and regarding a general intellectual curiosity as the prerogative of another class. When John Seller, a compass-maker at Wapping, sent magnetical observations to the Royal Society, he limited himself to commenting on what came 'within the Cognizance & sphere of my proffession and practice' as if he did not expect to be as broad-minded as the virtuosi, although he published the first adequate English pilot charts of the period and was appointed Royal Hydrographer.[52] Equally suggestive is Pepys' view of Richard Reeves, the premier optical instrument-maker of the day. For though

[48] See below, pp. 126, 138.
[49] Hunter,[107] p. 12; W. H. G. Armytage, 'The Royal Society and the Apothecaries, 1660–1722', *N.&R.R.S.*, 11 (1954), 24; Sloane 958, fol. 127 and passim; Hooke,[165] pp. 159, 208, 216, 267, etc.; Stearns.[151]
[50] Taylor,[156] pp. 221, 271; Tillison to Power, 23 April 1661, Sloane 1326, fols. 22v–23; Oldenburg to Boyle, 27 Oct. 1664, Oldenburg[121], II, 271; Hooke,[165] p. 267 and passim; R. W. Symonds, *Thomas Tompion* (London, 1951), ch. 2.
[51] Capp,[152] ch. 2, pp. 191–9; Aubrey,[161] I, 181–2; Sloane 2279–85; Bodleian Library MS Add. B 8; Taylor,[156] p. 241.
[52] Seller to Hooke, 12 April 1667, EL S.1, fol. 48; Taylor,[156] pp. 109, 244–5.

Reeves joined Pepys in using the instruments he bought and discussed related topics, 'it vexed me to understand no more from Reeves and his glasses touching the nature and reason of the several refractions of the several figured glasses, he understanding the acting part but not one bit the theory, nor can make anybody understand it – which is a strange dullness methinks'.[53] Both these men and their social betters felt that science was not really their business, due largely to their lack of education. The elementary matters that Coley had to explain to himself illustrate his shortcomings from this point of view, and Petiver was also an outsider to intellectual culture.[54]

Indeed here a great divide appears, with a clarity lacking in the relations between 'serious scientists' and virtuosi. For the educational background that these shared isolated them from those without it, and intellectual interests in men without proper schooling were frequently greeted with revealing expressions of surprise. The Oxford antiquary, Anthony Wood, thought John Graunt, tailor and pioneer in demography, 'the most ingenious person (considering his education and employment) that his time hath produced', learning like his being 'very rare in a trader or mechanic'. Similarly the excellence of the natural historical collection of the Bristol customs inspector, William Cole, seemed to the Oxford scientist, Edward Lhwyd, all the more remarkable for one 'who, perhaps, had not the advantage of a liberal education to invite him to such studies'.[55] Even the merchant Francis Lodwick – one of the few from that class who joined the Royal Society – encountered reservations when propounding linguistic reform on the grounds that he was 'indeed unequal to this undertaking, because deficient in learning, and born outside the academic sphere'. This applied to pursuits other than science, as with a gardener with antiquarian interests of whom it was said 'the want of meanes and learninge doth keepe him under'. Educational theorists made similar presumptions about the expectations of different classes in regarding learning as the prerogative of the wealthy while the poor were to train themselves to earn a living.[56]

Also revealing is the relation of such men to the coffee-houses

[53] Pepys,[163] VII, 254; Nicolson,[148] pp. 23–7.
[54] E.g. Sloane 2280, fol. 29v; Bodleian MS Add. B 8, fols. 28v–29; Stearns,[151] pp. 254–5.
[55] Anthony Wood, *Athenae Oxenienses*, ed. P. Bliss, I (London, 1813), 711–12; Lhwyd to Ray, 1 July 1690, Ray,[124] p. 225.
[56] Salmon,[211] p. 18n.; Douglas,[258] p. 259; Schlatter,[283] p. 50.

where scientific enthusiasts met. It was a commonplace that these
were patronised by people of diverse social backgrounds.

'Gentry, Tradesmen, all are welcome hither,
and may without Affront sit down Together',

ran a broadside expounding 'The Rules and Orders of the Coffee-
House', and in theory it should have been easy for a man of low
social class to learn about science there. But few can have afforded as
much time for endless talk as did Hooke and his virtuoso friends,
while John Houghton, a writer on agriculture and trade who will be
considered in the next chapter, thought that those best able to benefit
from coffee-house science already had a proper education.[57] However crucial, therefore, artisans' technical contribution to science –
as in the scientific instruments they produced – their lack of education
cut them and many merchants off from the centre of the subject more
completely than the virtuosi. The virtuosi represent the true social
extension of the science of the age.

The main centre of virtuoso culture was London, so it is not surprising that this was the seat of the Royal Society. Though the more
learned environment of Oxford and Cambridge encouraged the
proliferation of scientific as other intellectuals, contemporaries were
aware that London had much to offer that the universities lacked.
Coffee-houses abounded in London to an extent unparalleled elsewhere due to the sheer number of like-minded enthusiasts attracted
to the metropolis. Here books and curiosities were sold, collections
of rarities displayed and even experiments carried out, while the
discussions that went on are documented in Hooke's *Diary*.[58] More
personal groupings in the households of scientists and collectors
were also common. Edmund Wylde, a wealthy virtuoso and patron
of literati who lived in Bloomsbury, had a house described as 'a sort
of knick-knack-atory' where scientific enthusiasts frequently met to
view his relics, admire his excellent library and discuss matters of
common interest.[59] Another discussion group was presided over by

---

[57] Ellis,[149] frontispiece, pp. 44–6; Candill,[147] p. 369 and ch. 9 passim.
[58] Hooke,[165] pp. 85, 358 and passim; Caudill,[147] pp. 377–8.
[59] Roger North, *The lives of the Norths* (London, 1890), I, 374; [John Bagford], 'An
Account of several Libraries in and about *London*', *Monthly Miscellany*, 2 (1708),
179; Hooke,[165] p. 167 and passim; A. Powell, *John Aubrey and his friends* (London,
1948), pp. 255–6.

John Locke at Exeter House, the residence of Lord Ashley, who
in 1672 became Earl of Shaftesbury, and Shaftesbury (like other
aristocrats) also had a private laboratory.[60]

London could boast various facilities missing elsewhere. Due to
the Stationers' Company's tight monopoly in the book-trade, the
bulk of new works were published in London, and London booksellers were renowned for their excellent stock. It was also through
London that foreign books were imported, and, from the 1670s,
London became the centre of a trade in secondhand books sold at
auction. The ease of obtaining books was commonly seen as one of
the metropolis's chief advantages, and auctions and the catalogues
published in connection with them were particularly praised as 'a
great light to the knowledge of many rare Books'.[61] London was also
the source of all the 'new' scientific instruments such as microscopes
and barometers commercially produced, which were to be found
throughout England and even in Scotland.[62] It was also the chief
centre for the manufacture of more traditional engraved instruments
like quadrants, rules and sectors. As the country's commercial centre
and its largest port, demand focused on the metropolis and very few
instruments were professionally made elsewhere. In such matters
Oxford compared disappointingly: one commentator attributed the
relative poverty of scientific studies there compared with London
not least to the lack of instrument-makers and glass-grinders, which
made it difficult to 'well manage' experiments.[63] London could also
offer such miscellaneous facilities as libraries and botanical gardens
– notably those of the Apothecaries' Company and of Henry
Compton, Bishop of London – though these could also be found in
the universities.[64]

London, therefore, was the focal point of virtuoso culture, but
virtuosi were also widely dispersed throughout the country and even
beyond, into the colonies, where there was much curiosity about

[60] Cranston,[138] pp. 109, 117; K. H. D. Haley, *The first Earl of Shaftesbury* (Oxford, 1968), p. 220.
[61] Balle to Aubrey, 21 Aug. 1688, Aubrey 12, fol. 19; John Lawler, *Book auctions in England in the seventeenth century* (London, 1898); Bagford, 'An Account', p. 182.
[62] This is documented in the paper referred to in note 29, above.
[63] Taylor,[156] pp. 226, 255, 260; Bernard to Collins, 3 April 1671, Rigaud,[123] I, 159.
[64] Bagford, 'An Account', pp. 167–82; John But, 'Facilities for antiquarian study in the seventeenth century', *Essays and Studies*, 34 (1938) 64–79; C. Wall, H. C. Cameron and E. A. Underwood, *A History of the Worshipful Society of Apothecaries* (London, 1963), I, ch. 12; E. Carpenter, *The Protestant bishop* (London, 1956), pp. 366–7.

natural phenomena.[65] These scattered figures, too, played an important part in the science of the day: if anything, they were more significant than their London counterparts, providing invaluable observations for London scientists to consider and use as a basis for synthesis and theory. 'Though London be the seat of the wits, yet the country is the seminary', John Flamsteed claimed, apologising for adapting his astronomical observations 'to the meridian of a place no more famous than Derby'.[66] As shown in the last chapter, it was one of the accomplishments of the Royal Society, and especially of Robert Plot, to collect and orchestrate information from these scattered enthusiasts. Plot even hoped to use volunteers to collect a mass of data from which to write the natural history of all the counties in England, 'one of the most usefull & illustrious Workes that was ever produc'd in any age or nation' in John Evelyn's view, though Plot was overwhelmed by the undertaking and in fact only published accounts of two counties.[67]

Socially, the scattered enthusiasts for science in the provinces were very like those in London. The common educational background of the virtuosi provided a shared language and a network of communication transcending geographical distance: Plot seems to have used contacts with Oxford graduates in the counties. Any educated person could belong to the scientific scene, and some figures who lived far from the centres of science made theoretical contributions that were taken perfectly seriously, such as Henry Power or Thomas Strode in Somerset, considered by as eminent an expert as James Gregory 'very knowing in the mathematical sciences'.[68] It was also possible in the country as in London to accumulate libraries and 'repositories', of which perhaps the finest was that of the Leeds virtuoso, Ralph Thoresby.[69] But men with scientific interests had real problems in the provinces, due to the absence of the facilities which London so profusely provided, and the sheer isolation often felt without the gatherings of like-minded enthusiasts that were common in the metropolis.

Only through London contacts was it possible to get hold of scientific apparatus, and various protracted correspondences exist

[65] Stearns,[115] esp. chs. 4–7.
[66] Flamsteed to Brouncker, etc., 24 Nov. 1669, Rigaud,[123] II, 79.
[67] Plot, 'Plinius Anglicus', Society of Antiquaries of London MS 85; Gunther,[129] pp. 335–45; Evelyn,[164] IV, 68–9.
[68] Taylor,[156] p. 224. See also Rigaud,[123] II, 438–58.
[69] Thoresby, *Ducatus Leodiensis* (London, 1715), pp. 275–568.

on this subject between town and country virtuosi. As for books, the efficiency of the distribution network is striking, with a number of dealers throughout the provinces. The great difficulty, however, was in getting to hear of new books and knowing whether they were worth buying, 'theire being so many cheates in titles'.[70] This partly explains the appeal of periodicals like the *Philosophical Transactions* and the *Weekly Memorials for the Ingenious*. A scientist like John Ray also suffered in rural Essex from the lack of botanical gardens available in Oxford and London.[71]

Worst of all was the lack of scientific conversation, for though it was generally admitted that the country was to be preferred to London 'for the goodness of the place', London had clear advantages 'for the persons in it'.[72] This problem affected scientists and non-scientists alike. Philip Chester told John Aubrey that he found the country 'somewhat Melancholy, which makes me often wish my self att London, both for the diversion of the Town, & that I might enjoy the satisfaction of your Company, & the rest of my Friends there'. Some could commute to London for the season, but for those marooned in the 'Obscurity and Privacy' of their rural retreats the prospect was grim.[73] Letter-writing was only a partial substitute for contact. As Grew pointed out when writing to Lister via Oldenburg, an hour's conversation was preferable to a year's correspondence, in which it was difficult to answer the precise question troubling the other party. After retiring to Devonshire, William Balle urged Sir John Hoskyns to come and stay with him, if only for two or three days, 'wherein I can say more to you in a few houres then can bee written in many sheetes . . . you cannot imagine how itt will refresh mee'.[74] Cumulatively, it was possible for even 'a delicate inventive witt' to be thwarted for want of 'ingeniose conversation' and to get gradually more and more out of touch 'in an obscure corner'. Country virtuosi were more prone than London ones to become old-fashioned and eccentric in their views, humbly admitting that their ignorance of the latest scientific theories was greater than those

[70] See note 29 above; Johnston to Lister, 21 Jan. 1673, Lister 35, fol. 18; H. R. Plomer, *A dictionary of printers and booksellers . . . 1668–1725* (Oxford, 1922), pp. 329–42; Pollard,[160] pp. 9–17. [71] Raven,[135] pp. 193, 226–7, 294.

[72] Neile to Oldenburg, 5 Dec. 1667, Oldenburg,[121] IV, 8.

[73] Chester to Aubrey, 16 Oct. 1684[?], Aubrey 12, fol. 74; Hunter,[107] pp. 12, 58–9 (on commuting); Paschall to Musgrave, 3 Mar. 1685, Gunther,[129] p. 272.

[74] Grew to Oldenburg, 11 Dec. 1673, Oldenburg,[121] X, 396; Balle to Hoskyns, 3 July 1682, EL B.1, fol. 109A.

closer to centres of learning realised, while Hooke wrote off one such figure as 'a simple asse'.[75]

There were therefore strong incentives to form groupings for intellectual intercourse in the provinces, and it is indicative of the social environment of science that this happened most naturally in the meeting places of the leisured class. The earliest attempt to organise a group was in Somerset in 1669–70, when a '*Philosophical* Correspondence' was instituted as a means of 'bringing down among us what of these affairs is in London, and the two Universi-tyes, and consequently other where'. Its *Propositions* itemised its intentions: to collect information that 'may serve to promote, and inlarge the *History* of *Nature*', to find means for the hypotheses of members to be discussed both in the county and beyond it, and to provide scientific instruments and copies of the *Philosophical Trans-actions* and *Mercurius Librarius*, a useful broadsheet advertising recent books.[76] It is almost certainly significant that this took place around Bath and Bristol, which jointly supplied a background of leisure, government, the professions and commerce similar to that in London. There was also the comparable milieu of coffee-houses, in which Bristol was second only to London, while the kind of local audience to which science appealed is indicated by the complaints of the local collector, William Cole, about the idly curious lords and ladies who came to see his museum.[77]

It is equally revealing that of the societies successfully established in the 1680s, one was at Dublin, the second largest conurbation in the British Isles. Dublin was a social centre like London, as the seat of the English administration and the cultural and commercial capital of Ireland; here too, the group's original meetings were held in coffee-houses.[78] The other was more predictably at Oxford, but it is possible that this society – like its predecessor in the 1650s – is to be associated with the inroads made by coffee-house culture, of which conservatives complained. Elsewhere there were less formal meet-ings in the provincial capitals which increasingly attracted the local professional and leisured classes, including Norwich, York and

---

[75] Aubrey,[161] II, 169; Jessop to Lister, 3 Feb. 1674, Oldenburg,[121] X, 468; Hooke,[165] p. 292.

[76] *Propositions For the carrying on A Philosophical Correspondence, Already begun in the County of Sommerset, Upon incouragement given from the Royal Society* (London, 1670); Paschall to Glanvill, 18 June 1669, Oldenburg,[121] VI, 141.

[77] Ellis,[149] pp. 207f; Cole to Lister, 25 Nov. 1683, Lister 35, fol. 99.

[78] Hoppen,[117] p. 23.

Exeter, where a chemistry course was given in 1668 by Peter Stahl, who had formerly taught the subject in Oxford.[79]

The most fruitful local associations, however, were formed less by such spontaneous pressure than by scientists themselves, and these therefore reflected in microcosm the interrelationship between major researches and a wider group of enthusiasts that characterises the scientific scene. Such were the groups which surrounded Richard Towneley in Lancashire throughout the Restoration and that which – more briefly – formed around Martin Lister at York. These circles shared the scientific interests of their leading figures: Towneley's worked mainly on astronomy, meteorology and Galilean physics, in which it achieved much, even playing a part in the discovery and verification of Boyle's Law. The concerns of Lister's friends mirrored his preoccupation with natural history and the earth sciences, and physical science (for which Lister had little time) was barely represented.[80] The Oxford Philosophical Society was similarly moulded by its 'cheife' members, John Wallis and Robert Plot, while this tendency for groups to take up the interests of their leaders recurred in Somerset: the *Propositions* there made special provision for studying John Wilkins's 'Universal Character', thus clearly reflecting the enthusiasm of Andrew Paschall, a local divine who was one of the group's leading lights.[81]

Partly because of this dependence on individuals, however, such associations were precarious and subject to fairly rapid eclipse. The Somerset Correspondence only ran for eighteen months, though there was less formal collaboration thereafter, particularly on a project for writing the natural history of the county inspired by Plot in the 1680s.[82] Similarly, when Lister moved to London, the York circle faltered before developing its concerns in a different (less scientific) direction. Its limitations as a focus for local enthusiasm even in the 1670s are shown by the fact that a local gentleman, Richard Beaumont, elected to the Royal Society in 1684 and so likely

[79] Wood, *Life and times*, I, 201, 423, 472–3; Tuke to Evelyn, 28 Sept., 7 Oct. 1663, Evelyn, *Corr.*, nos. 1288–9; Frank,[40] p. 243; Peter Clark and Paul Slack, eds., *Crisis and order in English towns 1500–1700* (London, 1972), pp. 14–15.

[80] Webster,[158] pp. 65–73; Webster, 'Richard Towneley and Boyle's law', *Nature*, 197 (1963), 226–8; Carr,[142] esp. p. 24; Turner.[157]

[81] Wood, *Life and times*, III, 76–8; Gunther,[120] esp. pp. 45, 64; *Propositions*, p. 3; Salmon,[212] pp. 149–57.

[82] Paschall to Aubrey, 18 July 1684, Aubrey 13, fol. 67; Paschall to Musgrave, 3 Mar. 1685, Gunther,[129] pp. 272–3, and ibid, pp. 275–8.

to have had at least some interest in science, was evidently unknown to Lister.[83] For all the Royal Society's institutional weakness, comparison with these groups illustrates both its importance and the role of London in supporting activity on a scale unapproached elsewhere.

As the difficulties of organisation in the provinces show, it is easy to overestimate the popularity of science. The number of scattered scientific enthusiasts revealed by haphazard references sometimes seems surprising: for instance, Henry Power found 'many' instruments made by Richard Reeves in the north of England in 1662.[84] But, even so, these men were quite exceptional in their counties. In Somerset Sir Robert Southwell commented that he was not his neighbours' favourite, 'which I really congratulate, as much as I prefer Philosophy before drinking', while Joseph Glanvill complained how 'Our Gentry are of a temper very different from the Genius of the [Royal] Society'.[85] 'Nor could I amidst that variety of company wherewith our country abounds find any the least addicted to philosophy', complained Thomas Pigott, an Oxford don with scientific interests, when he stayed in Lancashire in 1678. John Ray was equally deprecating about his neighbours in Essex, and John Aubrey was inspired to write a play satirising the philistinism of his fellow countrymen in Wiltshire.[86] Such intellectual interests as were at all common in the counties were matters of immediate appeal to local landowners, like heraldry and antiquarian topography. Though scientists occasionally tried to exploit this to introduce more concern with natural history, it shows the problems the new science faced in achieving more than a minority appeal.[87]

Even in London many showed no interest in natural philosophy, despite all the evidence of a fashionable curiosity about science. This explains the common feeling among scientific enthusiasts that the response of their contemporaries was disappointing. In contrast to Sprat's assertion that experimenting suited the nation's 'present prevailing *Genius*', John Evelyn complained in 1679 of the Royal Society and its programme that 'Tis impossible to conceive, how so

---

[83] Hugh Honour, 'York virtuosi', *Apollo*, 65 (1957), 143–5; Lister to Oldenburg, 16 Nov. 1676, EL L.5, fol. 77.

[84] Power to Reeves, 10 Aug. 1662, Sloane 1326, fol. 31v.

[85] Southwell to Petty, 28 Nov. 1682, Lansdowne,[235] p. 113; Glanvill to Oldenburg, 15 Sept. 1668, Oldenburg,[121] v, 42.

[86] Pigott to Aubrey, 23 Feb. 1678, Aubrey 13, fol. 110; Ray to Lhwyd, 17 Aug. 1691, Ray to Aubrey, 4 July 1693, Ray,[125] pp. 181, 220; Aubrey,[161] II, 333–9; Hunter,[143] pp. 216–17.   [87] Laslett,[159] esp. p. 158; Hunter,[143] p. 70.

honest, and worthy a *design* should have found so few *Promoters*, and so cold a welcome in a *Nation* whose *eyes* are so wide open.' His remarks were echoed by other disappointed enthusiasts, not least John Beale, whose profuse letters to his scientific acquaintances return again and again to the need for concerted efforts to interest the 'debauched Gentry' in science.[88]

Such exhortations recur in the handbooks of virtuoso learning that streamed from the press throughout these years. Their very need to justify intellectual concerns in detail to all gentlemen shows their uphill struggle with more than a few. Typical in this respect is the plaintive defence of such matters in the preface to the publisher, Richard Blome's, compilation *The Gentleman's Recreation* (1686). Perhaps more revealing of the true avocation of the numerous members of the nobility and gentry who subscribed to its publication, however, was its actual emphasis (and especially the emphasis of its lavish plates) on outdoor sporting activities.[89] Furthermore, whereas such writers were thinking primarily of the privileged classes by whom science was chiefly patronised, the literate population of the country as a whole showed still less interest. Natural philosophy barely figures among the best-sellers of the day.[90]

On the whole, as is now appreciated, an appetite for science only became common in the eighteenth century. But popularisation began in the Restoration, as is shown best by changes in scientific instrument-making. In the first two decades of the Restoration such devices were almost entirely 'bespoke' from a few makers, who did not keep many items in stock: hence a customer often had to wait some time for his order to be completed. By the 1680s, however, this bespoke trade was supplemented by a more commercial one, with more craftsmen constructing instruments and a number of shops in London offering a wide selection for immediate sale to the public. They were now no longer 'rare and confined to the cabinets of the virtuosi' and only available through membership of virtuoso circles, as a contemporary noted.[91] The increased demand for instruments probably explains the introduction of techniques of mass-

[88] Evelyn,[197] sig. A3v. Cf., e.g., [Richard Graham, Viscount Preston], *Angliae Speculum Morale* (London, 1670), pp. 41–2; Hales, *Account*, p. xxxix; Beale to Evelyn, 3 May 1667, Evelyn, *Corr.*, no. 60 and passim.

[89] Blome, *The Gentlemans Recreation* (London, 1686), sigs. a1, a2v and passim.

[90] C. J. Sommerville, 'On the distribution of religious and occult literature in seventeenth-century England', *The Library*, 5th series 29 (1974), 224 and n.15.

[91] Goodison[155] pp. 30–4; Roger North, *Lives of the Norths*, I, 387–8.

production in lens-grinding and the rapid development of the screw barometer in the later years of the seventeenth century; the expansion of sales by the 1690s is clear from the appearance of advertisements for microscopes and telescopes even in almanacs.[92]

A related development was a growing market for purely popular scientific books in the last decades of the century. Perhaps the most famous was translated from the French, though it is striking that two English translations came out almost simultaneously in 1688: this was Bernard Le Bovier de Fontenelle's *Dialogues on the plurality of worlds*, which popularised the new cosmology in dialogue form. It was paralleled by William Leybourne's more lavish folio, *Pleasure with Profit* (1694), and other works which built on slighter precedents since the Elizabethan period.[93] Most significant and novel were the periodicals of the 1690s and especially the *Athenian Mercury*, which sought to disseminate knowledge of science and other subjects by answering questions sent in by readers. Its pages are revealing of both the prevailing ignorance in scientific matters and the patient and elementary exposition needed to overcome this: it was consciously aimed at a humbler social class than the Royal Society's publications, with lower educational expectations of its readers.[94]

Popularisation possibly had some adverse effects on scientific enterprise. The new technique of grinding groups of glasses together meant that requests for lenses of unusual thickness caused 'a grate deal of trouble', while Robert Hooke attributed the decline of microscopic studies in the late seventeenth century partly to the use of the microscope 'for Diversion and Pastime, and that by reason it is become a portable Instrument, and easy to be carried in one's Pocket'.[95] In contrast to the *Philosophical Transactions*, the *Athenian Mercury* had no place for original research, perhaps a reminder that Bacon condemned epitomes, despite the value he placed on

[92] R. S. Whipple, 'John Yarwell or the story of a trade card', *Annals of science*, 7 (1951), 67; Goodison,[155] ch. 3; Nicolson,[153] p. 6; M. H. Nicolson, *Science and imagination* (Ithaca, N.Y., 1956), p. 178.
[93] Mrs. A. Behn, trans., *A Discovery of New Worlds* (London, 1688); J. Glanvill, trans., *A Plurality of Worlds* (London, 1688); Meyer,[150] ch. 2; L. B. Wright, *Middle-class culture in Elizabethan England*, re-issue (London, 1964), ch. 15.
[94] Meyer,[150] pp. 49–59; W. Graham, *The beginnings of English literary periodicals* (New York, 1926), ch. 1; [Charles Gildon], *The History of the Athenian Society* (London,[1691]), p. 28.
[95] Whipple, 'John Yarwell', p. 67; Hooke, *Philosophical Experiments and Observations* (London, 1726), p. 261.

collaboration: his blue-print for science in the *New Atlantis* made no provision for popularisation.[96]

But criticisms like Hooke's are unfair. In the long run the extension of science's audience could only benefit research by providing a growing body of enthusiasts and a rising tide of popular acceptance. This laid the foundations for the eighteenth-and nineteenth-century triumph of science to which we are heirs, but which in Restoration England was in its infancy.

[96] Layton,[319] p. 10.

4

Utility and its problems

mmmmmmmmmmmmmmmmmmmmmmmmmmmmmmmmmmmmmmmmmmmmmmmmmmmmmmmmmmmmmmmmmmmmmmmmmm

Restoration scientists were obsessed by the usefulness of their studies. As Thomas Sprat put it in his *History of the Royal Society*, contrasting the new science with the sterile scholastic philosophy: 'While the Old could only bestow on us some barren Terms and Notions, the New shall impart to us the uses of all the *Creatures*, and shall inrich us with all the Benefits of *Fruitfulness* and *Plenty*.' In the Restoration, as in the Interregnum, many shared Bacon's conviction that the advancement of learning had suffered in the past not least through the dissociation of 'Speculative men' and 'men of experience': 'it were to be wished (as that which would make Learning indeed solid & fruitfull) that *active* men would or could become writers', so that natural philosophy might directly ameliorate human life.[1]

Just how literally such remarks are to be taken has been the subject of some controversy, partly because they contained an element of exaggeration which not all commentators have allowed for, partly because modern scholars have often let polemical preconceptions cloud their judgement, and partly because – on this as on so many methodological issues – there were differences of opinion at the time. At the most extreme, Sprat argued that it was for '*Mechanicks*, and *Articifers*' that 'the True *Natural Philosophy* should be principally intended'. Sprat's comments elsewhere in his book prove that he did not really hold as radical a view, but some evidently did. In the Interregnum Samuel Hartlib and his associates brought what can retrospectively be defined as 'science' and 'technology' so close together that it is difficult to disentangle them, particularly regarding priorities. Though not unimportant for encouraging scientists, Hartlib is most justly remembered for his attempts to apply scientific

---

[1] Sprat,[92] p. 438; Hunter,[143] p. 95.

knowledge to the needs of life: Charles Webster's *The Great In-staration* has cogently presented a continuum from theory to prac-tice in such subjects as chemistry, mineralogy and agriculture.[2]

Some scholars have taken similar arguments further, positing close links between scientific innovation and industrial and technical development throughout the early modern period. Bacon not only placed a high value on science's utilitarian benefits: he also held up the advancement of knowledge in the mechanical arts as a model to intellectuals, and in his high esteem of technical skills he had numer-ous sixteenth-century precursors, including the Spaniard J. L. Vives and the German Georgius Agricola. It has thus sometimes been claimed that the experimental method originated in the workshops of craftsmen, penetrating into natural philosophy to challenge and transform the impractical notions of the ancient Greeks. Edgar Zilsel placed in this context not only Bacon but also his precursor William Gilbert, whose *De Magnete* (1600) has often been seen as the true foundation of the empirical method in England.[3]

As a corollary, others have tried to prove that the priorities of scientific research were closely shaped by the technical and economic requirements of English society. R. K. Merton sought such evidence especially in mining and transportation, while the most extreme statement of this viewpoint came from Boris Hessen at the Inter-national Congress of the History of Science and Technology in London in 1931, when he itemised Newton's debt to such needs for the concepts that he considered in his *Principia*.[4]

Such views fail to do justice to the chief stimuli to scientific advance. Whatever the precocious empiricism of craftsmen's work-shops, the sixteenth century was significant for the realignment of intellectuals' priorities which allowed them to take artisans newly seriously. Similarly, Hessen's attempt to give primacy to external influences on the *Principia* represented a crude Marxist orthodoxy emanating from the zenith of self-consciousness of the newly-reformed Russian Academy of Sciences which few would now stand by. Even if Newton was aware of the potential relevance of some theorems of the *Principia* to practical needs (as occasional marginal

[2] Sprat,[92] pp. 117–18; Webster,[72] ch. 5 and passim.
[3] Zilsel, 'The origins of William Gilbert's scientific method', *J.H.I*, 2 (1941), 1–32; see also Paolo Rossi, *Francis Bacon: from magic to science* (English translation; London, 1968), pp. 1–11, and the studies listed below, p. 208.
[4] Merton,[168] chs. 7–8; Hessen,[167] pp. 155–76.

comments suggest), these applications were peripheral and the primary inspiration of his programme intellectual rather than utilitarian: it grew out of the predominantly explanatory problem-solving tradition deriving from Galileo, mentioned in chapter 1. 'I consider philosophy rather than arts', Newton explained, 'and write not concerning manual but natural powers.'[5]

The significance of the technological discussions in the Royal Society's minutes itemised by Merton has been challenged by a computerised analysis which found them a surprisingly minor component of the Society's organised activity. Indeed some contemporaries reacted against the overvaluation of utilitarian priorities in intellectual life that has been called 'vulgar Baconianism'. Thomas Hobbes complained how 'not every one that brings from beyond seas a new gin, or other jaunty device, is therefore a philosopher. For if you reckon that way, not only apothecaries and gardeners, but many other sorts of workmen, will put in for, and get the prize.'[6]

Hobbes was unusual (and unpopular in scientific circles) for his unrepentant elitism. But it is clear that most of those concerned with science had a strong sense of the essential primacy of intellectual considerations in the pursuit of knowledge about the natural world. Though anxious to see science applied, they were aware that too immediate a stress on practical considerations obscured the general principles on which these were based. This was Bacon's own view: despite his hopes for the amelioration of life and his belief in the fundamental unity of understanding with utility, he advocated a judicious balance of intellectual and practical ends, knowing that 'experiments of Light' were to be preferred to 'experiments of Fruit' when priorities clashed. As heirs to this principle Restoration scientists were truly Baconian, and even Sprat, though a more 'vulgar' Baconian than some, admitted 'that in so large, and so various an *Art* as this of *Experiments*, there are many degrees of usefulness'.[7]

There is a real danger of reading the appeal to utility too narrowly and assuming that it referred exclusively to practical, everyday needs. Many defenders of the utility of the Royal Society meant (like Glanvill in *Plus Ultra*) to protest the merits of improved knowledge of the natural world rather than utilitarian applications. The statistician John Graunt justified the Society's 'preparatory and luciferous

---

[5] Hall,[174] pp. 16–17; Werskey in Hessen,[167] pp. xiiif. Hall,[175] p. 8; Newton,[63] p. xvii.
[6] Frank,[106] p. 93 (cf. Hoppen,[117] p. 209); Purver,[24] p. xv; Hobbes,[70] IV, 437.
[7] Bacon,[69] IV, 17; Webster,[72] pp. 335–42; Rossi,[171] pp. 146–73; Sprat,[92] p. 245.

Experiments', associating the view that all experiments should have immediate value with its 'envious _Schismaticks_'.[8] The first and hence most prominent section of Robert Boyle's tract on this theme, _Some Considerations touching the Usefulnesse of Experimental Naturall Philosophy_ (1663–71), similarly dealt with the value of accurate knowledge about the natural world in its own right and for religious enlightenment, while another contemporary struck the right balance in acclaiming the work of the Royal Society as 'to the Glory of God, the promoting of true & reall knowledge & both the Profit & reputation of our Kingdoms'.[9]

It is even possible that the stress on utility had an element of public relations, emphasis on likely tangible benefits being intended to justify broader intellectual concerns to a hostile public. For it is essential to remember that the new science was attacked as trivial and unimportant, contrary to the claims of its propagandists. Prefaces to works by Fellows of the Royal Society often tried to answer the criticism 'What have they done?', and in 1666 there was a move to associate the Society with Sir Christopher Wren's plan for rebuilding the City of London after the great Fire just to rebut this charge, although the Society had played no part in the scheme.[10] Such arguments recurred later, as John Greenwood, a country correspondent of John Houghton's, explained:

I have often heard It obiected Against the Royall Society that though they are Ingenious men And have found out A great Many Seccrets In Nature yet what they have Done hath been Litle Advantage to the publick. Would they take Into Consideration the Nature of butter And cheese And prescribe Such Rules As might be plaine to the meanest Capacity I Am certaine it would fully Answer the obiection.[11]

This suggests artificiality, and the stimuli to intellectual activity are certainly distorted by overstressing practical considerations. But scientists nevertheless felt strongly that it was in their power to improve human life, either by applying theory or by bringing Baconian method to bear on industrial and other techniques. Some were more committed than others: not all felt as strongly as William Petty, who always warmly advocated utilitarianism, though perfectly

[8] Hull,[233] II, 323–4. [9] Boyle,[71] II, 5–63; DM 5, fol. 34.
[10] William Petyt, _Britannia Languens_ (1680) in J. R. McCulloch, ed., _A select collection of early English tracts on commerce_ (London, 1856), p. 357; Hooke,[64] sig. g1; Evelyn,[197] sig. A1v; William Petty, _The Discourse . . . Concerning the Use of Duplicate Proportion_ (London, 1674), pp. 1–3; Oldenburg to Boyle, 18 Sept. 1666, Oldenburg,[121] III, 231. [11] Greenwood to Houghton, n.d., Cl.P. xxv, fol. 133.

aware of the need to subordinate it to proper method. Moreover there were lesser men who did not aspire to intellectual innovation but participated in the technological programme: here the strong Baconian streak in Restoration science reappears, in works like Joseph Moxon's *Mechanick Exercises* (1677–84), which cited Bacon on the value of such investigations and their integrity to natural philosophy.[12] But these notions were strikingly central to the programme of most scientific thinkers.

They were particularly prominent in the early Royal Society, whose members felt strongly that a devotion to practicality should accompany the elucidation of natural effects. 'I shall not dare to think my self a true naturalist,' Boyle asserted, 'till my skill can make my garden yield better herbs and flowers, or my orchard better fruit, or my field better corn, or my dairy better cheese, than theirs that are strangers to physiology.'[13] Perhaps the best testimony to this is the project for a great collaborative 'History of Trades' that the Society espoused, which took up Bacon's notion of what was in effect the technological counterpart of his projected natural history. Information about technical processes was to be collected for its value in its own right and as a potential source of data for scientific hypotheses, while, through collation and comparison, it was also hoped that improvements noted in one area could be introduced in others. 'By this help the worst *Artificers* will be well instructed, by considering the *Methods,* and *Tools* of the best', Sprat believed, and it was widely felt that the inefficiency persisting in craft and industry could thus be eliminated through the efforts of intellectual overseers.[14] Boyle expounded such a programme in *Some Considerations . . .*, while Petty, Evelyn and Oldenburg listed trades suitable for study in this way.[15]

The History of Trades programme had creditable success in stimulating careful descriptions of industrial practices. The archives of the Royal Society are full of detailed accounts of all sorts of techniques – from the making of candles and parchment to that of cheese and pitch – though certain themes tended to be stressed.[16] One was dyeing, on which Petty compiled a paper which was published in

[12] Sharp,[144] ch. 1; Moxon, *Mechanick Exercises* (London, 1683), preface.
[13] Boyle,[71] II, 64.   [14] Houghton;[177] Sprat,[92] p. 310.
[15] Boyle,[71] III, 442–56; Lansdowne,[234] I, 205–7; Cl.P. III(1), fol. 1 (printed in A. F. Sieveking, 'Evelyn's "Circle of Mechanical Trades" ', *Transactions of the Newcomen Society,* 4 (1923–4), 40–7); Cl.P. XXIV, fol. 53.
[16] Cl.P. III(1), fols. 18, 22, 23 (printed in Birch,[118] I, 99–102), 26.

Sprat's *History of the Royal Society* and which is a model of precise analytical reporting; a special committee was formed to deal with cognate topics in 1666.[17] Another was tanning, the subject of much attention from the Honourable Charles Howard, brother of the Duke of Norfolk, who found a new way of making tannin without using bark: he advertised this in a printed broadsheet, evidently feeling (like others at the time) that acceptance of the novelty depended on such cheap publicity.[18] Other matters considered included salt-making, the production of alum and the brewing of cider, on which a large amount of research was published by the Society in 1664.[19] Mining technology also received extensive attention, not least in articles in early volumes of the *Philosophical Transactions*.[20] In each subject an attempt was made to check information and collate reports from different experts and hence to produce as authoritative an account as possible.

Such efforts had their agrarian counterpart in the concern to improve farming techniques of the so-called 'Georgical Committee', one of the groups set up to consider areas of the Society's interests in 1664. The intention of this was (in the words of a report of its meetings) 'the composing of a good History of Agriculture and Gardening, in order to improuve the practise thereof'.[21] It was hoped to examine carefully the agricultural methods of different counties, and to this end 'Heads of Enquiries', prepared after scrutinising relevant authorities, were circulated first in manuscript and then in the *Philosophical Transactions*. Many replies were expected, 'whereby, besides the aid which by this means will be given to the general End of collecting the aforementioned *History*, every place will be advantaged by the helps, that are found in any', and a number materialised. The committee also took an interest in horticultural

---

17 Sprat,[92] pp. 284–306; Sharp,[144] pp. 174–9; Birch,[118] II, 93, 97; Cl.P. vol. III(1), fols. 27–41; Sloane 852, fols. 1–7. See also Oldenburg to Boyle, 8 June 1666, Oldenburg,[121] III, 154; *Phil. Trans.*, 1 (1666), 362–3.

18 *Brief directions how to Tanne Leather according to a new Invention* (London, n.d.) (copy in Cl.P. vol. xxv, fol. 94); *Phil. Trans.*, 9 (1674), 93–6; Cl. P. III(1), fol. 12; Birch,[118] I, 387–8.

19 Salt: Cl.P. vI, fol. 20, x(3), fol. 13, xx, fols. 40, 42; *Phil. Trans.*, 4 (1669), 1025–8, 1060–7, 1077–9; alum: Cl.P. Ix(1), fol. 3 (printed in *Phil. Trans.*, 12 (1678), 1052–6); Royal Society MS Extra 1, item 2; cider: Cl.P. x(3), fols. 5, 14; 'Pomona, Or An *Appendix* Concerning Fruit-Trees' in Evelyn,[197] pp. 335–412.

20 *Phil. Trans.*, 1 (1665–6), 21–6, 45–6, 79–85, 330–43, 2 (1667), 525–7, 3 (1668), 767–71, 817–24, 4 (1669), 1080–3, 5 (1670), 1042–4, 1189–98, 1099–2002, 6 (1671), 2096–2113.   21 DM 5, fol. 65.

improvement, which received further attention from John Evelyn in his *Kalendarium Hortense* (1664) and other works.[22]

Related to this was a great enthusiasm for silviculture, and perhaps the epitome of this early collaborative activity was Evelyn's *Sylva, or a Discourse of Forest-Trees and the Propagation of Timber in His Majesties Dominions* (1664). This resulted from extensive researches by various Fellows of the Society in response to a request to help improve the timber supply from the Commissioners of the Navy. Though Evelyn's name is solely attached to the work, his primary task was to supervise the disparate work of others and to present it in an acceptable form with his 'exquisite pen'.[23] On its publication the book was a success, selling over a thousand copies in less than two years, 'a very *extraordinary* thing in *Volumes* of this bulk', as Evelyn was assured by booksellers. It reached a second edition in 1670 and a third in 1679, and by this time it was making a substantial profit for its publishers, which Evelyn regretted had not been diverted to the funds of the Society.[24] *Sylva* was a useful compendium of information on all aspects of planting and it inspired the accumulation of more, some of which was incorporated in successive editions.[25] More important, it certainly stimulated afforestation. Joseph Glanvill reported that local gentlemen to whom he had lent or given copies 'have to my knowledge been incited by it to plant some Thousands' of trees, and this was repeated widely. Indeed, there may have been truth in Evelyn's claim that the book was 'the occasion of propagating many millions of useful timber-trees throughout this nation'. The achievement was all the greater because (as was pointed out) it 'undertook the most difficult and hopeless of our Rural Improvements, for it requireth multitudes of hands at the work, and considerable Charges, and making the slowest return of Profit'.[26]

[22] *Phil. Trans.*, 1 (1665), 92; Lennard[179]; Denny,[178] pp. 483–5, 489–91.
[23] Newburgh to Croone, 31 Aug. 1663, EL N.1, fol. 1; Pett to Evelyn, 4 Nov. 1662, Evelyn, *Corr.*, no. 1086; Cl.P x(3), fols. 20–1, XIX, fol. 11; Sharp,[181] pp. 63–4.
[24] Evelyn, *Sylva* (London, 1670), sig. a1; Evelyn to Beale, 11 July 1679, Evelyn, *Lb.*, no. 406.
[25] E.g. Blount to Evelyn, 18 Jan. 1669, Christ Church, Oxford, Evelyn Collection, box of 'Loose letters to and from John Evelyn'; Walker to Evelyn, 15 Nov. 1700, 7 Feb. 1701, Evelyn, *Corr.*, no. 1339, Add. 15858, fol. 188; Sharp,[181] pp. 66–7.
[26] Glanvill to Evelyn, 15 Jan. 1671, Christ Church, Oxford, Evelyn Collection MS 3.2, fol. 10; cf., e.g., Aubrey 2, fol. 85, Dumaresque to Evelyn, 13 July 1670, Add. 15857, fol. 226; Evelyn to the Countess of Sunderland, 4 Aug. 1690, Evelyn[126], p. 317; *The True Domestick Intelligence*, no. 34 (31 Oct. 1679).

Much attention was also paid to new inventions, either propounded by the Society's Fellows or originating elsewhere but
brought to the Society for its approval and encouragement. In 1664
it was even suggested that the Society be given the duty of inspecting
all proposed mechanical devices to see if they were 'new, true, and
useful', a function later performed by the Académie des Sciences in
France.[27] Nothing came of the English plan, but ideas put to the
Society for comment ranged from an instrument for taking sound-
ings without a line to a superior type of candlestick.[28] Also, in 1664 a
committee dealing with 'mechanics' was set up – another of the
specialist groups inaugurated in that year – which discussed new
techniques in gunnery, nautical innovations and methods of raising
water, seeking in each to consider 'what hath been done in it hitherto, and whether and how it may be improved'.[29]

Equally important, intellectually fertile Fellows like Hooke and
Petty were encouraged to devise machines and gadgets, and not only
in the Society's earliest years but thereafter a number of fruitful
inventions came from scientific circles. In 1664 Hooke was asked to
produce technical aids 'acceptable and useful to the public' and he
obliged with suggestions for improving land carriage and 'a speedy
conveying of intelligence'. More significant were such devices as
Hooke's universal joint and 'wheel-cutting engine' and his contribution to improvements in horology like the spring-balance watch,
though there has been debate about the exact nature of his achievement in this field. Hooke was undoubtedly fertile in mechanical
inventions throughout this period, spending much time in experimenting on technical processes and in conversing with men
professionally engaged in them like the famous clock-maker Thomas
Tompion.[30]

Petty was especially interested in naval technology, and this preoccupied many Fellows. In the early 1660s Petty carried out extensive
trials with a prototype catamaran-style boat, *The Experiment,*
which Sprat punningly described as 'the most considerable *Experiment,* that has been made in this *Age* of *Experiments*', though the
Royal Society did not feel able to support the project financially. It is

[27] Birch,[118] I, 391 (cf. p. 116); Hahn,[99] pp. 22–3.
[28] RBO I, fol. 153; Cl.P. III(I), fol. 10.   [29] DM 5, fols. 66–7.
[30] Birch,[118] I, 379, 385, 463; 'Espinasse,[136] ch. 4; R. W. Symonds, *Thomas Tompion*
(London, 1951), chs. 3, 5; see also A. R. Hall, 'Horology and criticism: Robert
Hooke', *Studia Copernicana,* 16 (1978), 261–81.

less well known that in the same years the Society spent nearly £70 on an inventor called Roquefort who produced a new diving machine, though this was presumably a disappointment since nothing more is heard of it.[31] Prominent early Fellows such as Lord Brouncker, who was also a member of the Navy Board, tried to rationalise the techniques of shipbuilding. Brouncker claimed to have worked out 'exact rules for the building of a greater [ship] or less, and so make the whole business become a compleat and regular science; what is done now by the artists (as they are called) being merely by guess and custom'.[32] This subject also interested Boyle, as did the subsidiary naval matter of making drinking-water from sea-brine: Boyle was closely involved in promoting a device to do this on a commercial scale.[33]

There was also sustained interest in theoretical problems of practical significance – notably the longitude, attacks on which continued without any real success throughout the period. In 1662 a committee of the Royal Society examined a proposal of the navigational teacher, Henry Bond, for finding the position of a ship at sea from the variations of a magnetic needle. In the early 1670s another committee (not a Royal Society one, though various Fellows were on it) examined a second proposal by Bond and one by the Sieur de St Pierre, a protégé of Charles II's mistress, the Duchess of Portsmouth.[34] A related development was the establishment in 1675 of a Royal Observatory at Greenwich, where John Flamsteed became the first Astronomer Royal. The initiative for this came from Sir Jonas Moore, Surveyor of Ordnance, who had gathered around him at the Tower of London a group which showed a steady devotion to applied science. The intentions behind the observatory were single-mindedly practical: the royal warrant for its foundation began with the words, 'whereas, in order to the finding out of the longitude of places for perfecting navigation and astronomy, we have resolved to build a small observatory', and Charles saw its

---

[31]Sprat,[92] p. 240; Sharp,[144] pp. 245f; Royal Society accounts, 1661–3, 1664. See also Stimson,[286] p. 113; Cl.P. VI, fol. 28.
[32] Hill to Brooke, 12 Mar. 1663, Hill,[122] pp. 96–7. Cf. Evelyn[164] III, 327, 332–3; B. de Monconys, *Iournal des Voyages* (Lyons, 1666), II, 27.
[33] Sir P. Pett to Pepys, 3 May 1696, in Samuel Pepys, *Private correspondence and miscellaneous papers*, ed. J. R. Tanner (London, 1926), I, 115; Maddison,[140] pp. 147–57.
[34] Taylor,[156] pp. 102–3, 111–12; Bennett,[141] pp. 241f; Forbes,[183] pp. 8, 15–16, 17, 18–21; Hooke,[165] esp. pp. 96–7; Flamsteed,[162] pp. 37–8, 125–6.

justification in the need for precise data about the fixed stars 'for the use of his seamen'.[35]

Scientists also tried to purvey theoretical knowledge in a practical form. Perhaps the best example of this is Petty's *Discourse . . . Concerning the Use of Duplicate Proportion* (1674), intended not only to explain philosophical concepts 'in a way which the meanest Member of adult Mankind is capable of understanding' but also to promote the study of science 'by shewing the use of *Duplicate Proportions* in some of the most weighty of Humane affairs'. In it, Petty tried to bring Galilean physics to a wider public, illustrating the value of calculations in shipping, artillery and building. His was perhaps the most intellectually interesting of many efforts to teach applied science to the population at large, which also bore fruit in works like Moxon's *Tutor to Astronomie* (1659). At a more personal level, John Collins served as a broker between eminent scholars and more practical men needing theoretical assistance, persuading Newton and Gregory to provide formulae for calculating the capacity of wine-barrels which could be converted into tables for popular use.[36]

All in all, these efforts to find ways in which intellectuals could serve current needs show an impressive attempt at relevance. Underlying them is a concern to get to grips with the problems of the contemporary economy. The bid to describe and improve agricultural and industrial practices manifested a sense of the value of improved production. Timber supply was a real crux due to pressure from the combined demand of industry and shipping. Comparable importance could be attached to dyeing processes and salt extraction, while butter and cheese were 'the most considerablest manufacture In England'.[37] Above all, the close involvement of men of science in maritime affairs shows their conviction that intellectuals should contribute to this critical area of national life.

Furthermore, despite some claims to the contrary, a commitment to technology continued throughout the Restoration period. The corporate History of Trades programme may have proved a non-

---

[35] Taylor,[156] pp. 129, 233, 257, 269; Forbes,[183] p. 22 and ch. 1, passim; Flamsteed,[162] pp. xxv, 38, 39, 41.

[36] Petty, *The Discourse* . . . (London, 1674), sig. A4v and passim; Letwin,[216] pp. 107–8; Hall,[173] pp. 123–5.

[37] Wilson,[13] pp. 83, 186–7, 201 and ch. 7; Albion,[185] esp. ch. 3; Clow,[190] pp. 88–92; Sharp,[144] p. 174 (see also p. 205 and n.2); Greenwood to Maddocks, n.d., Cl. P. xxv, fol. 133.

starter and the georgical committee was barely active after 1665. But this was typical of the disappointment of early Baconian hopes for the Society's collaborative effort, and, in technology as in science, individuals compensated for this by working on their own. Metallurgy was surveyed by John Ray in the slightly surprising context of an appendix to his *Collection of English Words* (1674), while Martin Lister compiled a careful 'Methode for the Historie of Iron' in the 1670s and after, using information from a number of correspondents; one chapter of this was published in the *Philosophical Transactions* in 1693.[38] Salt-making also continued to arouse interest, as did other industrial techniques like cloth-dyeing and alum-processing.[39] A more miscellaneous concern for technological as well as scientific information is found in the natural histories that Robert Plot and John Aubrey wrote in the 1670s and 1680s. An interesting episode arose from this when Plot proposed changing the traditional method of barking and felling trees and hence improving the supply of timber to the navy, on the basis of his observations in Staffordshire.[40] The additions to successive editions of Evelyn's *Sylva* also show continuing interest in its subject.

The georgical committee found its heir in John Houghton, tea-dealer and entrepreneur. Unlike many of the earlier protagonists of technology, Houghton was little concerned with pure science except for conveying it simply to the public, claiming to 'greatly (although not altogether) avoid *speculation*, and chiefly mind those things that tend to *useful Practice*'. But he published a periodical, *A Collection of Letters for the Improvement of Husbandry and Trade,* full of detailed accounts of agricultural and technical processes, some of them by men (like Charles Howard and John Evelyn) who had collaborated in the earlier History of Trades. Houghton also reprinted the Royal Society's agricultural inquiries, to which he received several answers which he published in his journal.[41] There was equal continuity in maritime affairs. In the 1680s Petty

---

[38] Lennard,[179] pp. 27–8; Ray,[200] pp. 113f.; Lister 1; *Phil. Trans.*, 17 (1693), 695–9, 737–45, 865–70.

[39] Cl. P. vi, fols. 52, 54; *Phil. Trans.*, 12 (1678), 1059–64; John Collins, *Salt and Fishery* (London, 1682); JBO ix, fol. 167; Ray,[200] pp. 139f. Sharp,[144] p. 177.

[40] Plot,[67] ch. 9; Plot,[68] ch. 9; Hunter,[143] ch. 2; *Phil. Trans.*, 17 (1691), 455–61; Arthur Bryant, *Samuel Pepys: the saviour of the Navy* (Cambridge, 1938), pp. 235–6; Anthony Turner, 'Robert Plot on felling timber: the Baconian tradition in action' (unpublished paper read at Oxford in 1973).

[41] Houghton,[201] 1 (1692), no. 3; Houghton, *A Collection of Letters*, 1 (1681), 6–9, 127–36, 2 (1683), 75–9, 82.

remained as preoccupied as ever with naval technology, and both he and Hooke were encouraged in these years by Samuel Pepys, who retained a high view of the potential contribution of intellectuals to shipping.[42] Moreover Hooke remained no less fertile of inventions, devising a new windmill and an instrument for measuring the velocity of ships in the 1680s and further nautical gadgets in the 1690s.[43]

So arguments for a climax of activity in applied science in the early Restoration and a subsequent virtual abatement have been overstated: in fact, curiosity about technology characterised science throughout the period. But, even if exaggerated, there was some decline in such activity, and there were few successes to match that of *Sylva* in earlier years. Nothing came of Plot's proposal for timber-felling (though the government briefly took an interest in it), nor of most of the suggestions of scientists for changes in naval technology in the 1680s. It was increasingly clear that intellectuals could not effect improvements in practical matters as easily as they often supposed.

Moreover those concerned with applied knowledge – like Houghton – were increasingly isolated in their technological work even when frequenting scientific circles: Houghton was an active Fellow of the Royal Society, but his contributions seem almost out of place in its minutes. The younger generation of scientists was apparently less prone than the older to combine science and technology: it may be significant that by the 1680s John Aubrey felt bound to justify an article that John Flamsteed, the august Astronomer Royal, wrote for Houghton on malt-making (by the obvious argument that his father was a maltster), in a manner inconceivable in the atmosphere of the 1660s.[44] Similarly among lesser men 'curiosity' tended to gain at the expense of 'use', thus confirming the reservations about the virtuosi of some earlier critics. 'We amuse ourselves of what is wonderful, and think it below us to take notice of anything that is useful', wrote a commentator in 1698.[45] It is these phenomena that we must now try to explain.

[42] Lansdowne,[198] pp. 115f; Lansdowne,[235] pp. 119f; Rawlinson A 171, fols. 245–6; see also above, n. 40, and below, n. 62.
[43] Hooke, *Philosophical Experiments and Observations* (London, 1726), pp. 107–8, 225f; RBO VI, fols. 84–5; Deacon,[182] ch. 8.
[44] JBO IX–X, passim; Aubrey 2, fol. 104v.
[45] Prior to Lister, 19 Nov. 1698, *H.M.C. MSS of the Marquis of Bath*, III (1908), 294. Cf. Casaubon,[265] p. 31.

Of the reasons hitherto suggested for the decline in technology, perhaps most persuasive has been the argument that, after its hopeful beginnings at the Restoration, science became increasingly elitist between 1660 and 1700.[46] In a more extreme form, this might be put as the view that a commitment to technology was a legacy from the Interregnum which was not really in character with the new regime and which therefore steadily declined in importance. The argument is, however, faulty. The technological interests of Restoration scientists were not merely a continuation of those of the Interregnum. Some – like agriculture – were, and it has been cogently argued that *Sylva* owed much to work done by members of the Hartlib circle and other writers in the Interregnum and earlier. Other concerns, however, were quite novel. The strong interest in naval and military technology of Restoration scientists harmonised well with the nature of the new regime but it had little precedent in the Interregnum and is truly *sui generis*.[47]

Though aloofness undoubtedly existed, it was present from the start of the Restoration, balancing the impulse against it. The History of Trades was itself symptomatic of a realignment in European intellectual life, of the conviction growing since the Renaissance that intellectuals should involve themselves in practical matters traditionally considered beneath them: a gentleman was, after all, partly defined as a man who did not soil his hands by manual labour.[48] It is striking to find aristocrats and statesmen studying industrial processes in the early Royal Society and tempting to see this as a shift towards closer co-operation between intellect and economic life with wide implications. Yet the residual force of the older view is clear from Moxon's defence of such pursuits from elitist criticism in his *Mechanick Exercises*. The friction that resulted is in evidence at the birth of the History of Trades, in John Evelyn's complaint to Boyle in 1659 about 'the many subjections, which I cannot support, of conversing with mechanical capricious persons', and in the list of trades that he presented to the Royal Society: for he rather artificially

---

[46] 'Espinasse,[176] p. 74 and passim. For an alternative, less plausible, suggestion see Clark,[172] p. 17.

[47] Sharp,[181] pp. 64–5 and passim; Deacon,[182] pp. 73f; Hall,[173] pp. 61–2, 65–7, 119–20, 125. See Webster[72] for an apparent total lack of such interests in the Hartlib circle.

[48] William Harrison, *The Description of England*, ed. G. Edelen (New York, 1968), p. 114. See also Ruth Kelso, *The doctrine of the English gentleman in the sixteenth century* (University of Illinois studies in language and literature, 14 (1929)).

separated out those 'Usefull and purely Mechanic' from others 'Polite and More Liberall' and 'Exotick, & very rare Seacretts'.[49]

Such attitudes had a further corollary: intellectuals often valued 'secrets' on technological topics and were loath to vulgarise them by public dissemination. Much has been made of Bacon's attack on secrecy and his call for the free dispersal of knowledge: this assault on traditional attitudes which impeded scientific co-operation has rightly been seen as one of his most critical messages to the seventeenth century.[50] But the change in attitude that he advocated could not be effected overnight. Tension remained on the question, due partly to a virtuoso preoccupation with rarity, partly to the hard-headed business instinct of those who felt that they could profit by the innovations that they had recorded. John Evelyn's *Sculptura* (1662) – a history of the art of engraving – was intended as a contribution to the History of Trades, but he suppressed information about the method of mezzotint that Prince Rupert had taught him so that it would not 'be prostituted at so cheap a rate, as the more naked describing of it here, would too soon have expos'd it to'.[51] The problem was perhaps severest in subjects like shipbuilding where technical data could be regarded as arcana of state, but similar attitudes are to be found even in men like Boyle. Indeed, an ambivalence over whether to disseminate or withold knowledge presented a wider problem in the science of the time which deserves analysis.[52]

If secrecy negated the whole point of the History of Trades, however, elitism was not necessarily a bad thing. For, paradoxically, the technological programme was most successful when it was at its most discriminating. *Sylva* was a best-seller and greatly stimulated planting in the country, but it was widely criticised for its literary embellishments and difficult vocabulary, so that even if 'fittest for the Learnedst of our Nobility and Gentry', 'the countryman must go learn Latin and the poets to understand our author'. Hence calls

[49] Moxon,[199] p. 395; Evelyn to Boyle, 9 Aug. 1659, Evelyn,[126] p. 115; Cl.P. III(1), fol. 1 (cf. Christ Church, Oxford, Evelyn MS 65, front fly-leaf), with which it is instructive to compare Petty's classification in Lansdowne,[234] 1, 205–7, or Hooke's in Hooke,[65] pp. 24–6.

[50] Paolo Rossi, *Francis Bacon*, ch. 1; Houghton,[146] pp. 193f.

[51] C. F. Bell, ed., *Evelyn's Sculptura: with the unpublished second part* (Oxford, 1906), p. 148. I shall pursue this theme in my forthcoming study of John Evelyn.

[52] Hill to Brooke, 19 May 1663, Hill,[122] pp. 108–9; Aubrey,[161] II, 147; Hall,[76] p. 51; M. E. Rowbottom, 'The earliest published writing of Robert Boyle', *Annals of Science*, 6 (1950), 386; Maddison,[140] pp. 154, 165.

were made for a shorter and cheaper abridgement, while the third edition at least provided a glossary of certain difficult words.[53] Yet it was partly its refinement that ensured its success among the dilettante gentry, who were flattered by its pretensions. These passed their time, like Sir Robert Southwell when out of office, 'between Virgil's *Georgics* and Mr. Evelyn *On Trees*', and the book's excellent style – which impressed the Duchess of Newcastle – may well have helped 'to make gentlemen in love with the study'.[54] In any case, the squeamishness of some did not preclude different attitudes in others, and Robert Hooke was as happy with mechanics as intellectuals.

Obviously such fastidiousness did some harm, as the complaints about *Sylva* show. Similarly, the very complicated questionnaire circulated by the georgical committee was most suitable for the educated, and the replies to it – however retrospectively interesting – were very few considering the vast number of farmers in England. It was supplemented by a simpler rubric devised by Charles Howard which might have appealed more generally, though this never seems to have had wide circulation.[55] On the other hand, the success of *Sylva* may have indicated how improvements could be purveyed among the educated class in the provinces with which the Royal Society was in contact, whereas (as we shall see) influence on the more numerous lower social orders was much harder to gain. The possibility of using a network of prominent landowners to spread innovation seems to have occurred to the Royal Society's organisers, echoing a more general feeling that in agriculture the example of the socially elevated was a useful agent of change: this was shown by a proposal to encourage the use of potatoes in 1662.[56] The Society might even have been well advised to concentrate on relatively 'elitist' subjects.

But the whole question of elitism may be misconceived in implying that changing approaches to technology resulted from internal developments in the scientific community, rather than from the reaction of scientists to external factors that they met in trying to make

[53] *The True Domestick Intelligence*, no. 34 (31 Oct. 1679); Johnston to Slingsby, 20 Jan. 1666, *H.M.C. 6th Report (1877) Appendix*, p. 337; Pell to Haak, 30 Mar. 1667, Aubrey 13, fol. 93v; Walker to Evelyn, 15 Nov. 1700, Evelyn, *Corr.*, no. 1339; Evelyn,[197] sig. a2.
[54] Weld,[94] I, 324; Duchess of Newcastle to Evelyn, Feb. 1670, Evelyn,[126] p. 226; Johnston to Slingsby, 20 Jan. 1666, *H.M.C. 6th Report (1877) Appendix*, p. 337.
[55] Cl. P. xix, fol. 21; Lennard,[179] pp. 29–30n. and passim.
[56] DM 5, fol. 60; see also fols. 63–5 and below, pp. 108–9.

themselves useful. The most striking feature of their technological
programme was arguably the gap between ambition and achievement, and this most easily explains why attitudes changed as they
did, and why different people reacted in different ways. The aspirations of intellectuals to contribute to technical improvement have
often been cited as if effective in themselves, while the more difficult
but crucial question of what came of these hopes has been neglected.
What was attained, however, says far more about scientists' true
significance in this connection, as there is otherwise a danger of
attributing a central role to a peripheral phenomenon. Of all the
projects so far mentioned, however, few had more than slight effects,
and we must now consider why the results were so disappointing
compared with the good intentions we have chronicled.

The History of Trades and the projected account of agriculture
well illustrate the need to separate activity and results. Scientists
devoted much energy to such projects, but it is anachronistic to value
the information they recorded for its historical interest even when it
was never published. In its contemporary setting, the outcome was
less impressive. The enterprise was so vast in scale that contributions
towards it were bound to seem isolated and fragmented, even if this
was partly offset by concentrating on specific crafts. Just as revealing
is a contemporary's cry of frustration, in attempting a similar survey,
at the sheer range of informants needed to obtain accurate information on all aspects of agriculture and horticulture. Sir George Clark
sagely remarked how technology was simply less susceptible than
theoretical science to this systematic treatment, and even as
'materials for an encyclopaedia of early capitalist technology' – in
Clark's perceptive phrase – the Histories of Trades look patchy
and indifferent.[57]

As manifestations of this encyclopaedic impulse which was to
culminate in the eighteenth century they are not without significance, and they certainly helped teach intellectuals about practical
matters, which some saw as valuable at the time. Boyle considered
that such inquiry 'may enable gentlemen and scholars to converse
with tradesmen, and benefit themselves (and perhaps the tradesmen
too) by that conversation'. Similarly, in his 'Panificium' (an account
of the different ways in which bread was made in France), John
Evelyn thought it worth noting that stale bread could be freshened

[57] Richard Blome, *The Gentlemans Recreation* (London, 1686), sig. b1; Clark,[172]
pp. 28, 30.

by putting it in the oven, which was evidently news to his virtuoso audience at the Royal Society.[58] Even Henry Oldenburg had to copy quite elementary information from a nautical dictionary to prepare himself to comment about shipping, and there is similar evidence in John Aubrey's papers.[59] Purely descriptive accounts of practical processes at least had an educative value.

If the History of Trades was ever to serve the wider functions claimed for it, however, description and self-education had to give way to comparing different practices, to proving that some were superior to others and to championing and disseminating them. Even as serious a figure as Hooke, however, seems to have flagged at the tasks that this involved, though glad enough to outline techniques. He began his account of the felt-maker's trade by promising to draw inferences not only 'for the finding out of the operations of Nature' but also 'how this art may be varyed or improvd either as to the materiall, on which they work or as to the instruments & manner of their working, or both'. But he only ever completed the narrative part of the project, and others, having reached the same point, lamented the futility of heaping up mounds of 'tedious (because very ordinary) particulars'.[60] 'Thus have I made a long harvest of a little corne', the doctor William Jackson wearily concluded his account of cheese-making in Cheshire, while Silas Taylor complained of his work on cider: 'this I have done in obedience to your Comands; though it is like a Sermon at the Spittle onely repetition; for here nihil est dictum quod non fuit prius dictum [nothing is said that has not been said before]'. It is not surprising that so much of this inconclusive and miscellaneous material remained in manuscript in the Royal Society's archives to be discovered by historians, unless it was published to fill space in the *Philosophical Transactions* in 1678 when more up-to-date articles were in short supply.[61]

To transcend such inconclusiveness a more practical stimulus was perhaps required than the generalised curiosity underlying these reports. Though information could be efficiently processed and applied to more intellectual questions, as shown by the collaborative work on tides mentioned in the last chapter, the technological coun-

[58] Boyle,[71] III, 401; cf. Sharp,[144] p. 172; Houghton, *A Collection of Letters*, 1 (1681), 135.

[59] Cl. P. VII(1), fol. 20; Hunter,[143] p. 110 n. 12.

[60] Cl. P. XX, fol. 96; Tenison to Oldenburg, 7 Nov. 1671, Oldenburg,[121] VIII, 348.

[61] Cl. P. III(1), fol. 22; RBO II, 192 (this sentence is omitted from the printed version in Evelyn,[197] pp. 397–400); *Phil. Trans.*, 12 (1678), 931–6, 945–52, 1046–71.

terpart to this was too often lacking. It is arguably significant that the queries of the Navy Commissioners inspired the sustained research that went into *Sylva*, and the navy was a centralised organisation where suggestions for change could be immediately applied throughout the period. Indeed Samuel Pepys, the supreme rationaliser of the Restoration navy, himself frequented scientific circles and his papers contain numerous projects for maritime improvements. It is as yet difficult to assess the result of this, but there are hints that even Pepys lacked the application to bring such schemes to fruition, taking Petty less seriously than he would have liked.[62] In France effective and single-minded deployment of similar piles of information was illustrated by the chief minister, J. B. Colbert's, attempts thus to improve techniques, as in his regulations for the cloth industry. Though English intellectuals expressed interest in the information that these contained, however, they were not in a position to follow Colbert's example for reasons which will be examined in the next chapter.[63]

More serious was the fact that though intellectuals supposed that they could easily and quickly master any craft and suggest improvements overlooked by ignorant artisans, this was not necessarily the case. It is most revealing that intellectuals assumed that they could achieve so much so easily, and one can learn much about the cultural topography of the day by reading their disdainful comments about rude mechanics. Even Hooke sometimes showed himself scornful of 'the words or reports of some cosening workmen'.[64] Yet it early became obvious that many processes were not easily described in words or even diagrams, since the operation had to be carried out precisely to be successful: 'the practick part' was just not susceptible to this kind of treatment.[65]

Matters were also often more complicated than they at first seemed, particularly to people newly approaching the subject. Hence, even when suggestions for improvement were made, they were frequently less useful than they appeared. Deeper inquiry often

[62] Rawlinson A 171 and subsequent vols., passim; Pepysian 977, 1074, 1731, 2501, 2910; Petty to Southwell, 23 Dec. 1684, Petty to Deane and Pepys, 30 Dec. 1684, Lansdowne,[198] pp. 140–1, 149–50.
[63] King,[229] pp. 210–11 and chs. 5–7 passim; William Aglionby to Sir J. Clayton, 5 July 1671, Cambridge University Library MS Dd.11.57, fol. 86.
[64] Cl. P. xx, item 50, fol. 106.
[65] Deane to Pepys, 27 Oct. 1668, Rawlinson A 195, fol. 60; cf. Birch,[118] 1, 116; C. M. Cipolla, 'The diffusion of innovations in early modern Europe', *Comparative studies in society and history*, 14 (1972), 48, 51.

suggested that divergent operations were not due to ignorance but more complex factors. The Royal Society's agricultural inquiries evidently indicated the variety of farming practice caused by the conditions in different regions which modern scholars have particularised. This taught at least some Fellows (such as Sir John Hoskyns) that it could not be assumed that techniques successful in one area would work elsewhere, and it may explain why the inquiries were never followed up.[66] Similarly, Pepys – although curious about experiments in such matters – noted the 'folly' of the view that one nation was better than another at building ships or anything else: for the contrasts among houses around the world were due not to greater or lesser knowledge but to diverse circumstances.[67]

Inventions entailed comparable problems, since, despite the optimistic salesmanship of their protagonists (often echoed in modern books), they were not always as feasible as was claimed. In the Interregnum, Petty's instrument for 'double-writing' evidently encountered obstacles not least because it did not work very well, and similar snags hindered the machines for making salt-water sweet by which Boyle set such store after the Restoration.[68] The scepticism of contemporaries was well expressed by the Anglican cleric, Brian Duppa, in a letter to Sir Justinian Isham, about another project: 'The truth is I look'd upon it, as you do in your mathematicks upon many conclusions, which in the theory are specious [i.e., plausible], but when they are reduced to practice do many times fail us'.[69] Opinions differed about the feasibility of many hopeful schemes and devices – from a plan for a canal linking the Thames and the Bristol Avon to Petty's 'double-bottomed' ship, which aroused widespread reservations although it is possible that it might have worked.[70] Equally contradictory were views on the advisability of certain agricultural practices, while Pepys's judgement of a calculating machine designed by the free-lance inventor, Sir Samuel Morland, was 'very pretty, but

---

[66] Joan Thirsk, ed., *The agrarian history of England and Wales, IV, 1500–1640* (Cambridge, 1967), ch. 1; Kerridge,[194] esp. ch. 2; JBO IX, fols. 237–8.

[67] Pepys,[237] p. 166. See also Pepys,[236] pp. 240, 351–2.

[68] Sharp,[144] pp. 238–9; Maddison,[140] p. 156.

[69] Duppa to Isham, 16 June 1657, in Sir G. Isham, ed., *The correspondence of Bishop Brian Duppa and Sir Justinian Isham 1650–1660* (Northamptonshire Record Society, 1955), p. 136.

[70] John Aubrey, *The Natural History of Wiltshire* (London, 1847), pp. 30–2; Long to Aubrey, 5 Mar. 1683, Aubrey 12, fol. 279; Lansdowne,[198] pp. 119–20 and passim; Sharp,[144] pp. 245–6.

not very useful'.[71] This naturally dampened the enthusiasm of those who financed such projects: one can understand the Duke of York's resentment about unsuccessful experiments with a new type of shot in 1667, whose inventor 'as tis said by some hath spent his majestie in practices of this Nature £1100'.[72]

Intellectuals were also bad at judging economic factors, though sophistication grew during the Restoration, so that in Houghton's writings improving proposals were backed up with thorough arguments only rarely paralleled earlier. In the overpopulated agrarian economy of seventeenth-century England there was little need for the labour-saving gadgets of scientific enthusiasts, so they were never taken up. This explains why nothing came of John Evelyn's suggestion that the 'sembrador' – a Spanish machine that ploughed, sowed at an equal depth and harrowed all at once – be introduced into England, whatever its success in as depopulated a country as Spain.[73] Lack of demand probably also accounts for the failure of Charles Howard's mode of making tannin without bark. The price of bark was falling rather than rising so there was little incentive to adopt his method, which was slower than that usually employed since tannin had to be extracted from the branches where it was less concentrated.[74]

Another difficulty arose in naval matters. Even when the theoretical knowledge of intellectuals gave them an advantage over others – as in doing experiments on fluid displacement and making calculations about the most efficient form of hull design – the technical problems involved were too great for them to master. The results were therefore disappointing, even if they sometimes challenged accepted views.[75] Though Sir Peter Pett, a member of a famous shipbuilding family, asked Sir Christopher Wren for advice on ship design, giving him information and assistance, this proved 'all in vaine as to any illustration of the affairs of naval architecture that

[71] Greenwood to Maddocks, n.d., Cl. P. xxv, fol. 133; Maddocks to Greenwood, 28 Sept. 1685, ibid, xxv, fol. 134; Pepys,[163] IX, 116–17; see also H. W. Dickinson, *Sir Samuel Morland* (Cambridge, 1970), pp. 28–33.

[72] Bodham to Pepys, 26 July 1667, Rawlinson A 195, fol. 114.

[73] Houghton[201] and *A Collection of Letters* (1681–3), passim; D. C. Coleman, 'Labour in the English economy of the seventeenth century', *Economic History Review*, 2nd series 8 (1956), 280–95; Clark,[172] pp. 45f, esp. p. 48 n.1; *Phil. Trans.*, 5 (1670), 1055–65.

[74] I am grateful to Dr L. A. Clarkson for this information, communicated by personal letter. See above, p. 92.

[75] Sharp,[144] pp. 242–3, 246–7; Rawlinson A 341, fols. 25–7.

Sir Christopher could effect'.[76] It was very difficult to combine the intellectual's understanding and the artisan's practice, and, faced with the choice, the rulers of the navy preferred the views of shipbuilders with much pragmatic knowledge but little scientific training, such as Sir Anthony Deane and Peter Pett (Sir Peter's cousin). Their distrust of scholarly assistants was voiced by the engineer, Sir Henry Sheeres, who argued the need for experienced seamen, not theoreticians: 'to Discourse speculatively on points Practicable and of Demonstracon is like a blind Man's treating about Colours'. These reservations are echoed in Pepys's jottings in his naval journals, doubtless after discussions with such men.[77]

One should therefore treat sceptically the favourite suggestion of intellectuals as to why their proposals were not generally adopted – the ignorance and conservatism of craftsmen, shipwrights and farmers. It was claimed at the Royal Society that the 'sembrador' failed because 'people are so attached to their old way, as to neglect the use, though more beneficial, of a new one', and this was widely echoed. 'I have sed as much as this 100 tymes: but I find it hard to convince this Age that they should be wiser than their fore fathers', wrote another disappointed improver.[78] It was also claimed that the procrastination and 'interest' of 'Envious *Plebeian* Mechanicks' thwarted innovations in maritime affairs like the introduction of milled lead for sheathing the hulls of ships.[79] 'We live in an age wherein men are more apt to mock a man if unsuccessful than to commend what is successfull, & much lesse to contribute', complained the inventor Gilbert Clerke – a useful caveat to Defoe's optimism in his *Essay upon Projects* about 'the general Projecting Humour of the Nation' – and similar comments intersperse Petty's reports of his shipping experiments.[80] But, in view of the impracticality of some proposals and the differing opinions on the feasibility of others, one sympathises with those farmers who were inclined to write off the suggestions of the educated as the 'whimsyes of contemplative persons', while Boyle, for one, was an earnest apologist for the good sense of

---

[76] Pett to Pepys, 3 May 1696, in Pepys, *Private correspondence*, I, 115–16; Bennett,[141] pp. 232f, 267.

[77] Rawlinson A 341, fol. 25; Pepys,[236] e.g. pp. 16, 158, 228.

[78] Birch,[118] II, 425; Cl. P. xxv, fol. 120.

[79] JBO x, fol. 96; Hales, *Account*, pp. vi, viii, xxi and passim.

[80] Clerke to [?], 7 Nov. 1688, Cl. P. III(1), fol. 67; Defoe,[202] p. ii; Lansdowne,[198] passim.

artisans. There was also a longstanding tradition of (not unjustified) hostility to projectors.[81]

In this, there was truth on both sides. On occasion the diffusion of novelty was undoubtedly impeded by conservatism, interest, and the secrecy of workmen who feared losing the advantage they derived from unusual methods (though scientists were not blameless on the same score, as we have seen).[82] Despite this, however, the findings of economic historians do not suggest a general opposition to technical innovation, and scientists' most extreme denunciations must be rejected. The received picture of the agricultural history of the era — based on the record of men like Aubrey — is of a steadily accelerating acceptance of improvements like the use of root crops and new fodders and the growing of specialised cash crops in areas in which they were economically advantageous. In industry there were comparable developments: it has even been argued that the habit of undertaking small-scale adjustments necessitated by the replacement of wood by coal (itself a crucial adaptation) gave English artisans the attitudes which helped bring about the Industrial Revolution.[83] On the whole the interest of intellectuals like those of the Royal Society followed rather than led such changes. Modifications in processing alum and copperas were both merely reported to the Society, while the agricultural inquiries naturally revealed innovation rather than led to it. Even when intellectuals made investigations these were often irrelevant to practice, so that improved techniques noted by travellers abroad were never applied although desperately needed at home.[84]

Just how much books by scientists and others contributed to technical change is uncertain, and, though research is in progress on the agricultural side of the question, it is doubtful if the evidence will permit a firm answer.[85] Current indications are ambivalent. In agri-

---

[81] Smith to Beale, 18 Sept. 1655, EL S.1, fol. 2; Boyle,[71] III, 442; Wilson,[13] pp. 101–3; Clark,[172] pp. 104f.

[82] E.g. Deane to Pepys, 27 Oct 1668, Rawlinson A 195, fol. 60; Dummer to Pepys, 1 Feb 1679, Rawlinson A 172, fol. 26; J.B.O. IX, fols. 177, 179, 211–12. See above, p. 100.

[83] Thirsk and Cooper,[17] pp. 177–9, and see the studies cited on p. 209, below; Clark,[172] pp. 43–5; Harris,[189] pp. 4–9.

[84] *Phil. Trans.*, 12 (1678), 1058–9; Royal Society Ms Ex 1, item 2, e.g. fol. [13] (evidently by Sir Hugh Cholmley: see DM 5, fol. 77); Lennard,[179] e.g. pp. 34, 44; Hollister-Short,[191] pp. 160, 162, 169–70.

[85] Joan Thirsk is investigating this matter for the forthcoming vol. v of the *Cambridge agrarian history of England and Wales*.

culture, men of high social rank showed a disproportionate tendency to innovate, while treatises like John Worlidge's *Systema Agriculturae* (1669) were undoubtedly aimed at the gentry, with literary embellishments for their benefit like Evelyn's in *Sylva*. But a recent study of improvement in Wales suggests that, though books played their part, they were only effective in conjunction with other, more personal forms of influence which satisfied farmers of the feasibility of novelties. Probably capital rather than science explains why men of higher rank were more prepared to take risks.[86] In industry, the most serious claim for a bookish stimulus to invention concerns an educated man, Thomas Ravencroft, who devised a practicable form of lead-containing glass. But the influence of a printed treatise even on Ravencroft and certainly more widely is open to doubt, since it was reprinted only on the continent and much of its novel terminology never caught on. Change generally occurred at the level of artisans and entrepreneurs, and books were probably largely irrelevant to processes that were often very complex and hard to describe. Furthermore, as Boyle observed, 'it is not the custom of tradesmen to buy books', and the literacy rate of craftsmen was much lower than that of their social betters.[87]

On the other hand, a market undoubtedly existed among yeomen farmers and practitioners of mechanical arts for handbooks describing techniques and disseminating theoretical skills like arithmetic, a market which more general natural philosophy never reached. To such readers John Houghton deliberately appealed (in contrast to the more aloof approach of *Sylva* and the agricultural inquiries), publishing 'useful things fit for the *Understanding* of a plain Man' in cheap, easily available instalments, so that 'not only the Theoretical Gentleman, but also the Practical Rustic may enjoy their benefits'.[88] Houghton was unusual in his close association with the Royal Society and this audience was mainly served by men with expert knowledge of their chosen topics. They included Edmund Bushnell, author of *The Compleat Ship-Wright* (1664); James Lambert, who wrote *The Country-Man's Treasure* (1676); Samuel Sturmy, whose *Mariners Magazine* (1669) was stored with 'Mathematical and Prac-

[86] E.g. Wilson,[13] pp. 152–4; Thirsk,[192] pp. 153–4; John Worlidge, *Systema Agriculturae* (London, 1669), preface and passim; Emery,[196] esp. pp. 47–8.

[87] Turner,[180] pp. 206, 208, 212–13; W. A. Thorp, *English glass*, third edition (London, 1961), pp. 143–60; Harris,[189] esp. p. 8. Boyle,[71] III, 397; D. Cressy, 'Levels of illiteracy in England, 1530–1730', *Historical Journal*, 20 (1977), 5–6.

[88] Houghton,[201] 1 (1692), no. 1; Houghton, *A Collection of Letters*, 1 (1681), 3.

tical Arts of my own practice'; and Joseph Moxon, who produced a whole series of manuals of which his *Mechanick Exercises* is perhaps most notable. Such books give detailed and reliable information, clearly expressed, about a whole range of crafts and skills: as such they have been of immense value to historians, while the popularity of at least some of them is attested by the frequency with which they were reprinted.[89]

But the impact even of these books was limited, particularly when it came to suggesting changes rather than chronicling existing practices for the benefit of novices. Their authors were often as critical as intellectuals of the conservatism of their peers, and among both them and scientists frustration gave rise to the conviction that the novelties that they humbly suggested to their readers, backed by plausible arguments about feasibility, could be more widely and effectively disseminated by legislation.[90] Hartlib had thought in the Interregnum that improvement should not 'be left wholly to the uncertaine, disorderly and lazy undertakings of private men'. He was echoed by John Greenwood in 1685, who told Houghton that he had 'made soe many vaine Attempts to perswade country people out of theare Comon Rode that I Am Apt to Conclude that all you can say or Doe will signifie Litle without the Assistance of Authority'.[91] The political significance of this attitude will be assessed in the next chapter, but it illustrates clearly the difficulty of disseminating new techniques. The laborious and unrewarding tasks of the literary purveyor of innovation discouraged many who were less self-effacing and public-spirited than Hartlib in the Interregnum and Houghton in a later period.

The large and diffuse audience for such information and the slowness with which it could be reached – let alone change effected – contrasted markedly with the tightly-knit European intellectual community. Here the efficiency of networks of communication like Oldenburg's correspondence popularised scientific discoveries and theories almost overnight, ensuring swift and widespread acclaim

[89] Sturmy, *The Mariners Magazine* (London, 1669), sig. a1; Donald Wing, *Short-title catalogue of books printed in England . . . 1641–1700* (New York, 1945–51), s.v. 'Bushnell, Edmund', 'Moxon, Joseph'. On mathematical books see Taylor.[156]

[90] E.g. Andrew Yarranton, *The Improvement improved*, second edition (London, 1663), sigs. A3v–4; John Wilcox to Hooke, 15 Sept. 1681, Sloane 1039, fol. 105v; Plot,[68] p. 382.

[91] Hartlib, *Legacie* (London, 1651), sig. A2; Greenwood to Houghton, 14 Sept. 1685, Cl. P. xxv, fol. 132.

for their inventors. The experiments in blood transfusion at Oxford and London in the late 1660s were quickly reported at home and abroad: so quickly, indeed, that parallel attempts were made in Paris, and a typically acrimonious dispute over priority ensued. Moreover inventions by English virtuosi like Morland's 'Loud-Speaking Trumpet' may have been better-known in the international scientific community than in England due to their enthusiastic promotion by Oldenburg.[92]

It is therefore not surprising that scientific discovery had attractions for intellectuals that technological improvement lacked, resulting in an inevitable preference for theoretical over practical pursuits. It is easy to oversimplify this process, since the utilitarian Baconian preoccupations of science meant that the two were never exclusive. But this, more than anything, explains the Royal Society's withdrawal from technological improvement, while the growing complexity of scientific knowledge also encouraged this dissociation to an extent not precedented in the Interregnum. Intellectual life was satisfying whereas, both now and earlier, attempts to improve technology ended in repeated disappointment.

It is hence revealing that scientists who aspired to a more significant role than that of professor did not go to technology to find it. Instead, they moved out of science into positions of power where they could use their talents in the more traditional world of public affairs. A classic paradigm of this was the Interregnum career of Petty. Petty had been attracted to the Hartlib circle in the late 1640s by the opportunities it offered him as a scientist and inventor. Yet he was clearly soon disappointed and, in the 1650s, he moved away, first to academic life, and then to serve the state through the Down Survey. This employed 'scientific' concepts and an ability to organise on a daunting scale to achieve a landmark in the accurate assessment of resources for planning purposes: it might have been the beginning of a promising career in administration but for complications to be examined in the next chapter.[93] In the Restoration other scientists moved away from intellectual affairs as they became ambitious for more influence: Newton left Cambridge to be first Warden and then Master of the Mint, while the supreme example was Sir Christopher Wren. After a brilliant career as an astronomer in the 1650s, Wren clearly became dissatisfied with what science and academic life could

---

[92] Oldenburg,[121] IV, xx–xxi and passim; VIII–IX, passim; *Phil. Trans.*, 6 (1672), 3056–8.   [93] Sharp,[144] ch. 1; Strauss,[217] parts 1–2; see above, p. 25.

offer, and instead devoted his fertile intellect to the Surveyorship of Works.[94]

These were particular solutions to the problem presented by the desire of scientists for influence beyond the realm of the intellect. But the role they found arguably owed most to exhortations to intellectuals to pursue the public good that go back to the so-called Commonwealth writers of the sixteenth century: it had less to do with Bacon's utilitarian programme, which proved, for many reasons, so disappointing.

[94] Manuel,[139] ch. 11; Bennett,[141] ch. 6; Sir J. Summerson, *Sir Christopher Wren* (London, 1953), pp. 60–1.

5

Politics and reform

Can the new science in seventeenth-century England be associated with any particular religious or political ideology? Of all aspects of science's social relations, this has perhaps attracted most attention, but it is a topic fraught with difficulty. The great danger is of trying to prove too much and ending by proving too little, of attempting to link science with a particular sect or party but defining the members of this so broadly that only platitude or special pleading results.

This is clearly illustrated by perhaps the most famous of all such essays, the argument that Puritanism was especially conducive to the rise of science in early modern England. This thesis originated in the 1930s and was interesting as a precursor of modern attempts to integrate science with its context at a time when few showed interest in such an exercise. But its limitations are now all too obvious, and it might have been quietly abandoned but for its recent revival by Christopher Hill and others. Even Charles Webster includes some incautious claims of this kind in *The Great Instauration*, though they are not borne out by the profuse information in his book.[1] It is therefore worth indicating some of the problems involved.

Major claims for Puritanism are often based on a confusion of definition. In his classic exposition of the links between science and Puritanism, R. K. Merton used 'Puritan' to mean virtually the whole of early modern English Protestantism, thereby proving a generalised connection between religious and intellectual trends which is indisputable because it is so vague. In contrast, the more common usage of Puritanism associates the term only with the more precisian wing of the English church whose oppositionist role in the Civil War has earned it the name 'the Puritan Revolution'. Putative relations

[1] Webster,[72] esp. ch. 6. For a survey of other writings on this subject see below, p. 210.

between this and the new philosophy present far more difficulties. It is notoriously hard to isolate this Puritan wing of the church even in religious terms, since it had a constant tendency to merge back into the broader Anglican tradition. This problem is compounded when correlations are sought in non-religious matters, since ingenuity can extend the category of Puritan almost indefinitely: it is thus not clear that an associate of Hartlib's like Petty can be seen as a religious Puritan in any meaningful sense, though he has sometimes been placed in this category.[2]

It is certainly wrong to see such Puritanism as promoting science in a way that other ideologies did not. On the contrary, there is good evidence that Puritanism in its precisian sense was not specially favourable to natural philosophy. One of the few specifically Puritan societies in the seventeenth century – colonial New England – had no extraordinary commitment to science, and this is equally true of the most characteristic, clerical wing of English Puritanism. It was said that Presbyterians forbade the study of the new philosophy at some Oxford and Cambridge colleges in the Interregnum, and a recent study has discerned hostility among strict Puritans to the arguments for the religious value of the study of nature of which (as we saw in chapter 1) scientists made so much.[3] It has been argued that 'revolutionary' Puritanism after 1640 was *sui generis* in attracting to its ranks those committed to intellectual change, its characteristics proving nothing about Puritanism more generally.[4] But even then the emergence of political and religious radicalism in the revolutionary years and the concomitant reaction add further complexities, some of which were indicated in chapter 1. Puritans, like others, may have contributed to scientific enterprise, but Puritanism can claim no preponderant share of its achievement.

Some of the problems of the links between science and Puritanism also arise with the most popular candidate for its replacement, Latitudinarianism, particularly Latitudinarianism with the political dimension for which recent exponents have argued. As mentioned in

---

[2] Merton,[168] p. 57 and chs. 4–6 passim; on the difficulties of 'Puritanism' see especially J. W. Allen, *English political thought 1603–44* (London, 1938), part IV; Basil Hall, 'Puritanism: the definitional problem', *Studies in church history*, 2 (1965), 283–96; C. H. George, 'Puritanism as history and historiography', *Past and Present*, 41 (1968), 77–104.

[3] Perry Miller, *The New England mind: the seventeenth century* (Cambridge, Mass., 1939), ch. 8; Patrick,[91] pp. 22–3; Morgan,[208] esp. pp. 553–60.

[4] Rabb,[206] pp. 63–6.

chapter 1, the Latitudinarians were that party at the Restoration who reacted against the sectarianism of the Interregnum, advocating a simple theology and a tolerant church; they also espoused a political and economic viewpoint which justified the social order of Restoration England and its mercantile potential. The integrity of Latitudinarianism to the Interregnum ideology of the new science has already been indicated, and there was a marked overlap in personnel between the Latitudinarians and leading scientists and scientific enthusiasts in the Restoration, including churchmen like John Wilkins and Joseph Glanvill and laymen like Sir Robert Moray and George, Baron Berkeley.[5]

But even this connection of science and religion needs qualification, for, though it contains an element of truth, it is in danger of proving so much that it is platitudinous. Latitudinarianism, like Puritanism, is notoriously difficult to define, and we have seen how Latitudinarian ideology was championed by the propagandists of the early Royal Society to draw together all reasonable men and smooth over divisions by stressing what all had in common. It is hence not surprising that an attempt has been made to revise the Latitudinarianism thesis in the direction of religious neutrality or even indifference, though this has not been altogether convincing.[6] The reaction to sectarianism placed a natural premium on the advocacy of moderation and restraint and the promotion of the national good: but this might only disguise the remaining differences between royalists and parliamentarians, Whigs and Tories, Anglicans and Dissenters, High Churchmen and Low Churchmen.

What is most striking about the religious and political affiliations of those devoted to the new science – after the Restoration as before – is their heterogeneity. As Charles Webster has demonstrated, the active nucleus of the Royal Society in its earliest years was divided almost evenly between men who had collaborated with the parliamentary regime in the Interregnum – Jonathan Goddard, William Petty and John Wilkins – and its intransigent opponents – John Evelyn, Sir Robert Moray and Christopher Wren. This can be paralleled throughout the period. One might attach significance to the enthusiasm for science of a churchman like John Wilkins – the

---

[5] Shapiro,[85] pp. 153f, 226. See also Jacob,[307] ch. 1, and the studies listed on p. 203, below.

[6] Mulligan,[210] pp. 110–11; Mulligan, 'Anglicanism, Latitudinarianism and science in seventeenth-century England', *Annals of science*, 30 (1973), 213–19.

epitome of Latitudinarianism and a vociferous proponent of comprehension, who came to the climax of his career and even the prospect of major political advancement when hopes for such a policy were at their height at the end of the 1660s. But, if so, he should be balanced by Seth Ward, Bishop of Salisbury, his opponent on toleration, the author of an influential sermon advocating passive obedience, *Against Resistance of Lawful Powers* (1661), who also took a close interest in the Royal Society and its affairs after the Restoration.[7]

Similarly, if John Ray or the mathematician John Pell are evidence of the 'left-wing' affiliations of science, Thomas Willis and Isaac Barrow had clear connections with the 'right'. This is true of the younger generation as well as the older, in the contrast between two medical scientists, Nehemiah Grew and Martin Lister. Grew was a committed dissenter and a warm advocate of 'occasional conformity', the formal church-attendance to avoid penalties that was anathema to High Churchmen. Lister, on the other hand, was a 'stiff and conservative' figure, a strong champion of hierarchic values who mistrusted innovations both within medicine and without. Equally, in the physical sciences, the Tory Edmond Halley balances the Whig Isaac Newton.[8] This is true of minor enthusiasts as well as major scientists, for it is certainly wrong to argue that serious scientists were inclined to the 'left' while the 'right' supplied amateurs. Edmund Wylde, who appeared in chapter 3 as the epitome of a dilettante virtuoso (though one of some scientific significance), had been a Member of Parliament associated with the Interregnum regime, while the Suffolk doctor, Nathaniel Fairfax, a classic inconsequential fact-gatherer, was a nonconformist minister.[9] These are just as typical of their kind as such better-known royalist and Anglican dilettantes as John Evelyn or Joshua Childrey.

In the Interregnum some links have been suggested between scientific interests and attitudes and religious and political viewpoints, but these are easily overestimated, as we have seen.[10] Both then and after the Restoration, science had a genuine neutrality. Individuals

---

[7] Webster,[72] p. 95; Shapiro,[85] pp. 169f.; W. G. Simon, 'Comprehension in the age of Charles II', *Church History*, 31 (1962), 440–8; Thomas in Nuttall and Chadwick[10] pp. 195–206.

[8] [Nehemiah Grew], 'The True Catholick', Sloane 1941, fols. 7–12; Carr,[142] pp. 58–9, 384–5, 388; Halley,[238] pp. 268–9.

[9] Anthony Powell, *John Aubrey and his friends* (London, 1948), p. 256 n.1; Aubrey,[161] I, 251, II, 142; Oldenburg,[121] III, xxiv, 321n and passim.

[10] See above, pp. 26–7.

had very different motivations, but no convincing evidence has been adduced to link these exclusively with either the subject or method of their scientific research. Moreover, though some had strong affiliations of one kind or another, these are balanced by many more who were only marginally, if at all, committed on ideological issues. It is thus very difficult to find more than the vaguest hints about Hooke's political and religious views from his *Diary*, in contrast to that of the more transparent Evelyn. Such considerations further complicate attempts to find specific ideological connections for science, and this, more than anything, undermines faith in the quantitative analyses that have been made of the loyalties of Fellows of the Royal Society. These have also, however, often been vitiated by an uncritical counting of heads, regardless of the differing significance of more or less active sections of the membership, and hence (as with other essays in quantitative history) have provided a perfect excuse for those who want to dismiss their conclusions.[11]

But this need not induce complete pyrrhonism about the search for affiliations for the new science. So far as any correlation has been significant, its proof has not been a precarious demonstration of statistical predominance among scientists but a genuine unity of attitude in scientific and non-scientific matters. If 'Puritanism' is defined to include all but extremists among English Protestants, there is no doubt that commonplace attitudes very loosely defined as 'Puritanical' were favourable to the growth of science (an alternative suggestion that a so-called 'hedonistic-libertarian ethic' was crucial is less convincing, even if containing a grain of truth).[12] Similarly, the links between science and Latitudinarian ideas were significant not only at the time of the Restoration but thereafter, as is revealed not least by the hostility that they jointly encountered.[13]

This may be paralleled in politics, for one source of evidence about science's political affiliations is often neglected. The actual practice of scientists is less revealing than their ambitions, their hope that the principles and methods of the new philosophy, if applied to the economy and society, would bring about improvements of wide importance. Scientists wished to bring rationalisation and order to all areas of national life, and, though they would often have denied it, this had political implications in a contemporary context. Even if more marked in some than others, this connection retains its signifi-

[11] Feuer,[209] ch. 2 and appendix C; Mulligan[210]; for a critique see Webster,[205] pp. 20–1.
[12] Feuer,[209] ch. 2.   [13] See below, chs. 6–7.

cance. The aspirations of some enthusiasts were used to tar the new science with a single brush, rather as natural philosophy was sometimes suspected (despite the diversity of its supporters' opinions) of being 'a Lattitudinarian designe to propagate new notions in divinity'.[14] There is even occasional evidence that the two attitudes went together, though the political views of the Latitudinarians have been little explored.

The clue to this mentality will be found in the reformist plans that flourished in scientific circles, plans which are worth surveying in their own right, not least since their vogue in the Interregnum might wrongly suggest that they died out at the Restoration. There were some significant shifts in priority, and a tendency, hardly surprising in the current mood of reaction, to suppress the innovations of the revolutionary regime. But, despite this, the new science was associated, after 1660 as before, with a strikingly ambitious frame of mind, which was certainly not the prerogative of a single religious or political faction. Even the millenarianism which had accompanied such hopes during the Puritan Revolution was echoed in a purely Royalist setting in the 1660s and was clearly widely diffused. Indeed, though attention has been drawn to the almost apocalyptic expectations for science in Sprat's *History of the Royal Society* as evidence of the residual influence of Puritanism, this is mistaken, for such ideas were part of the general intellectual equipment of the age.[15] Even if some in scientific circles were more concerned with reform than others – as during the Interregnum – yet the extent of interest in major reorganisation is striking.

This is partly shown by the schemes for technological improvement surveyed in the last chapter, both in intellectuals' frustration at what they more or less justly saw as the inefficiency characteristic of technical processes and in their hopes to remove it. These were all too rarely realised, and the rather limited success of aspirations to economic and technological reform was typical of the difficulties encountered more generally.

More arresting, however, were the projects for a new, 'universal' language based on rational principles, which would supersede the 'Defects, and æquivocations' of existing tongues. Such schemes originated in the Interregnum, but they reached their climax in John

[14] Glanvill to Oldenburg, 17 Dec. 1669, Oldenburg,[121] VI, 372.
[15] M. McKeon, *Poetry and politics in Restoration England* (Cambridge, Mass., 1975), ch. 8; Webster,[205] p. 22.

Wilkins's classic book, *An Essay towards a Real Character, and a Philosophical Language* (1668), which epitomised the hopes of such thinkers. The intention was to produce a clear and systematic language that could not only be used by merchants, divines and scientists but would also objectively classify ideas and data about natural phenomena in what might 'prove the shortest and plainest way for the attainment of real Knowledge, that hath been yet offered to the World'.[16] Wilkins shared his interest with many others, including Seth Ward, who made a similar proposal, though its emphasis differed, being more concerned with logic and less with classification and communication; in the 1670s, after Wilkins's death, a group of like-minded enthusiasts tried to improve and refine his system. Though, like other grandiose schemes of the time, they succumbed to technical problems, they testify to the concern for fundamental change associated with science.

Less spectacular but equally significant was a committee established at the Royal Society to consider how the English language could be regulated and improved by the supervision of a group of intellectuals, after the manner of the Académie Française in France. Little is known of the few meetings that took place, but John Evelyn wrote a letter to Sir Peter Wyche, their convenor, in which he advocated the compilation of a proper grammar, the standardisation of orthography, and the supervision of vocabulary: 'there ought to be a law, as well as a liberty in this particular'.[17] This concern for uniformity may partly have inspired the interest in dialect of scientists like Ray, while a related impulse underlay the Royal Society's concern for 'plain style', of which so much has been made by literary historians. This reflected a similar wish for clarity, but the role of science against other factors has in the past been much overestimated, since this stylistic change was connected with more general shifts in the intellectual climate of seventeenth-century England, of which the growth of science was only one.[18]

---

[16] Hunter,[143] pp. 219, 60–3; Wilkins,[232] sig. b1v; Salmon,[211] esp. chs. 2–4; Salmon[212]; Knowlson,[213] ch. 3.

[17] Evelyn to Wyche, 20 June 1665, Evelyn,[126] pp. 159–62; see also Evelyn to Pepys, 12 Aug. 1689, ibid, pp. 309–11.

[18] Salmon,[211] pp. 102–3; R. F. Jones, 'Science and English prose style in the third quarter of the seventeenth century' in *The seventeenth century: studies in the history of English thought and literature from Bacon to Pope by R. F. Jones and others writing in his honor* (Stanford, 1951), pp. 75–110; George Williamson, *The Senecan amble* (London, 1951), esp. ch. 9; see also Cope,[137] ch. 7; Shapiro,[85] pp. 74f.

The same quest for order and improvement is also evident in the investigation by Restoration scientists of a possible universal measure based on the arc of swing of a pendulum or a degree of the earth's latitude. This engaged the attention of Hooke and others, and John Graunt cited it when dedicating his *Natural and Political Observations . . . upon the Bills of Mortality* to the Royal Society in 1662 to illustrate the importance and potential utility of the Society's work.[19] Such a unit of measurement would have had intellectual and practical benefits, helping solve the problem of longitude which preoccupied so many and improving standards of mensuration at a more humble level. It was hoped thus to reduce the fraud which flourished because of the widespread diversity in weights and measures and so help the poor, though, as John Aubrey found, 'what the reason is I cannot tell; but the poor (who are ignorant) are against it'.[20]

Aubrey illustrates another interest of the virtuosi, educational reform. Again, this was based on a dislike of obscurantism and an appeal to simplicity, clarity and efficiency. As in other reformist projects, Interregnum precedents were consolidated, with the crucial exception that the social engineering in at least some schemes of the Hartlib circle was now dropped. Instead, the emphasis was on improving the education of the ruling class, though technical education for essential occupations, such as sea-officers, was not neglected. It was also characteristic of the period that such schemes as came to fruition owed nothing to public patronage, apart from the Royal Mathematical School attached to Christ's Hospital: the field was dominated by private educational establishments, which were not only meant for dissenters excluded from the universities but were patronised even by the aristocracy.[21] None of these survived long, but the hopes of the time are as significant as their rather disappointing outcome, for in their challenge to the techniques and content of traditional education they were quite as radical as their predecessors: Aubrey's unrealised plans in his *Idea of Education* illustrate this clearly.[22]

---

[19] Hooke,[65] pp. 458f; Cl. P. II, fol. 3, xx, fol. 34; Bennett,[141] pp. 256f; L. D. Patterson, 'Pendulums of Wren and Hooke', *Osiris*, 10 (1952), 288–9, 313–14; Hull,[233] II, 324.   [20] Hunter,[143] p. 219.

[21] Hunter,[143] pp. 50–5; Webster,[72] pp. 207–17; Webster,[89] passim; Vincent,[214] pp. 93f, 199f; Caudill,[147] pp. 319f, esp. p. 328. On the Royal Mathematical School see below, pp. 132–3.

[22] Aubrey 10; the edition by J. E. Stephens, *Aubrey on education* (London, 1972) is unfortunately highly defective.

Most significant of all was the 'political arithmetic' pioneered in the Restoration by Sir William Petty, building on his experience in the Down Survey of Ireland in the 1650s, which can be seen as the application to social and economic affairs of techniques and ideas borrowed from natural science. 'God send mee the use of things, and notions, whose foundations are sense & the superstructures mathematicall reasoning', Petty wrote, 'for want of which props so many Governments doe reel & stagger, and crush the honest subjects that live under them'. He intended 'to express my self in Terms of *Number, Weight,* or *Measure*' rather than 'only comparative and superlative Words, and intellectual Arguments', seeing such factual precision as essential to efficient government (he also wished objectively to refute emotive views that the nation was decaying).[23] Precise data was to be collected and processed to form the basis of policy and of the planning and regulation of the national economy: a typical scheme was for a 'Register Generall of People, Plantations, & Trade of England', intended to 'give the King a true State of the Nation at all times'. Petty's friend, John Graunt, similarly justified statistical data of the kind that he presented in his *Natural and Political Observations . . . upon the Bills of Mortality* (1662), 'by the knowledge whereof, Trade and Government may be made more certain and Regular'.[24] Moreover in his writings, Petty moved on from retailing information to make prudential comments and rationalising observations about the state of the country and how it could be improved by efficient use of the available resources.

In this there is the germ of an approach to government which has become increasingly common since. Petty has been rightly celebrated for the acuteness of his economic analysis, despite his tendency to tamper with his figures to prove what he wanted: an extraordinary range of concepts was first enunciated by him, including the labour theory of value.[25] What is significant here, however, is the manipulative intention which Petty made explicit in his writings, his view that 'the Impediments to *Englands* greatness, are but contingent and removable'.[26] This was implicit to a greater or lesser extent in all the plans for the rationalisation of national affairs that have been mentioned – from language to industry – and it reveals a clear political

[23] Lansdowne,[234] I, 111; Hull,[233] I, 244; Sharp,[144] pp. 122–36, ch. 4; Letwin,[216] ch. 5.
[24] Lansdowne,[234] I, 171–2 (cf. ibid, sects. 4, 7 passim); Hull,[233] II, 396.
[25] G. Routh, *The origin of economic ideas* (London, 1975), pp. 35–46; Letwin,[216] pp. 134–7.   [26] Hull,[233] I, 298.

dimension to such schemes, rendering a conclusion about science's affiliations inescapable.

Such views are clearest among Petty's published works in his *Treatise of Taxes* (1662), which interspersed proposals for an equitable and efficient taxation system with reflections on related themes with obvious autocratic implications. Not only would the poor be set to work, but the numbers entering the universities and the professions would be moderated. Petty saw the latter as monopolistic and inefficient: he hoped to limit the number of doctors according to statistics of illnesses and to save resources by reducing the number of parish priests and even reintroducing celibacy.[27] Such ideas are echoed in Petty's working papers, unpublished till this century. When considering the rebuilding of London, he revealingly speculated on what might be done 'supposeinge all the ground and Rubish were some one man's who had ready money enough to carry on the worke, together with a Legislative power to cut all Knots'. For Ireland, he went on from advocating a register of lands, commodities and inhabitants to argue that the knowledge of the country thus obtained could be used to make its agriculture and industry more efficient: 'Wee can see whether plenty makes them lazey, and remeddy it', he noted, hoping to adjust the number of merchants to that of manufacturers and that of mariners to merchants. Petty's disregard for individual interests opposed to the corporate national good is clearest in his enthusiasm for large-scale transplantation of population: at one time he considered moving the bulk of the native population of Ireland to England and replacing them with Englishmen.[28]

This distaste for inefficiency and this appeal to the authority of the central government can be paralleled in other scientists who thought about the wider implications of their studies. Thus Nehemiah Grew wrote a truly remarkable tract on 'The Meanes of a most Ample Encrease of The Wealth and Strength of England in a Few Years'. Though dated 1707, this draws on ideas of this period, advocating the planning of resources, the introduction of practical education, the imposition of economic improvement even on those who resisted it, and still more extreme interference by the state to forcibly curb the idleness of the gentry. Such ideas are reflected more widely in the conviction of many technological writers that the best way to imple-

[27] Hull,[233] I, 23–5, 27–8, 29–30.
[28] Lansdowne,[234] I, 28, 90 and 77f passim, 46–7 and sects. 3, 10 passim.

ment their innovations was by backing them up with government legislation: as we have seen, the feeling that people should be compelled to mend their ways is as widely encountered among Restoration projectors as among Interregnum ones like Hartlib.[29]

In political terms, scientists' concern to extend efficient planning to all areas of national life aligned them with a 'strong' government which alone could overcome the resistance of sectional interests. In the Interregnum they naturally appealed to the republican government, and the profuse efforts of Hartlib and his friends to achieve national prosperity in co-operation with the regime are perhaps better seen as evidence of this ambition than of 'Puritanism'. Then in the early Restoration attention turned to monarchic government, and, in the 1690s, to the effective parliamentary executive established by the constitutional settlement following the Revolution of 1688. Whatever the regime's political colouring, however, these enthusiasts for reform and improvement always tended to approve the strengthening of the executive at the expense of impediments to its efficiency.

It is hence hardly surprising that both scientists and enthusiasts for technological improvement envied the power of the state in contemporary France and covertly wished that it could be emulated in England. Thomas Hales was typical in his respect for 'that mighty spirit of Industry' which ensured that a canal could be driven from sea to sea in France – 'a work that heretofore abash'd the *Roman Empire*' – while in England it was even impossible to prevent encroachments on the Thames. Representative also was his citation of 'the common *Observation*, that *Prerogative in the Hand of the Prince is a Scepter of Gold, but in the Hand of the Subject a Rod of Iron*'.[30] Samuel Pepys similarly lionised France, while Henry Oldenburg went further in speculating on how much contemporary northern monarchs could achieve if they were as powerful as their predecessors in the ancient Near East. It is worth recalling here that Bacon's *New Atlantis* was a totalitarian state, and Bacon's political attitudes had much in common with those of his Restoration successors.[31]

---

[29] British Library MS Lansdowne 691; see Johnson,[220] ch. 7; see above, p. 110.

[30] Hales, *Account*, pp. cxxi, lvif., cxiv.

[31] Pepys,[236] passim; *Phil. Trans.*, 6 (1671), 2091; T. K. Rabb, 'Francis Bacon and the reform of society' in T. K. Rabb and J. E. Seigel, eds., *Action and conviction in early modern Europe* (Princeton, 1969), pp. 169–93.

Petty and others claimed that they were above politics, that the pursuit of the national good should convince all sensible men to leave their wranglings. As John Beale put it during the severest constitutional crisis in Charles II's reign, 'the veryest trifle that is a reall Improvement may doe more good, than all those mighty matters which begot the greatest heats in Parliament ever since my Memory'.[32] It is symptomatic of the same frame of mind that John Collins – himself a keen protagonist of enlightened government interference in economic affairs – should have believed that mathematical insight could ease the relations between the English government and the seditious planters of New England. Petty was more forthright still, claiming that his improvements in County Kerry were effected 'without passion or Interest faction or party, but as I thinke according to the Aeternal Lawes & measures of Truth'.[33]

These sentiments, however, were ingenuous or misconceived. Restoration England was torn by a bitter conflict between executive and legislature that has often been seen as the key to the political character of the period, and in this context ideas like Petty's had clear and immediate political implications, while Pepys's and Hales's admiration of France would have appalled many. It is not surprising that Petty was an advocate of the standing army that terrified the parliamentary gentry, and his sympathies were all with the central government. Though allowing a slightly reluctant place for the consent of parliament on relevant issues, he felt that stability was essential and a strong executive its best guarantee.[34] Indeed, Petty saw parliamentary sovereignty and constitutional conflict as an obstacle to national success: to quote his exact words, 'a second Impediment to the greatness of *England*, is the different Understanding of several Material Points, *viz.* Of the Kings Prerogative, Privileges of Parliament, the obscure differences between Law and Equity; as also between Civil and Ecclesiastical Jurisdictions.' He was therefore keen to have these difficulties removed, seeing none as 'Natural'. It is interesting that he thought that the changes that he hoped for would be more feasible in 'poor Ireland' than in 'proud England'.[35]

---

[32] Lansdowne,[234] I, 103; Hull,[233] II, 395–6; Beale to Evelyn, 5 May 1680, Evelyn, *Corr.*, no. 144.

[33] Letwin,[216] ch. 4; Collins to Beale, 20 Aug. 1672, Rigaud,[123] I, 203; see below, p. 125; Sharp,[144] p. 210.

[34] Lansdowne,[198] p. 45 (cf. L. G. Schwoerer, '*No standing armies!' The anti-army ideology in seventeenth-century England* (Baltimore, 1974)); Hull,[233] II, 630–2; Lansdowne,[234] I, sect. 1, passim; W. H. Greenleaf, *Order, empiricism and politics* (London, 1964), pp. 255–6.   [35] Hull,[233] I, 300, 301; Lansdowne,[234] I, 108.

The political implications of rational planning are equally clear from the work of Petty's friend, Sir Peter Pett, whose *The Happy Future State of England* (1688) justified statistics in phrases like Petty's, arguing that 'the knowledge whereof is the Substratum of all political measures that can be taken as to a Nations strength or riches'. For Pett – who thought that he had proved the impossibility of a Catholic restoration – was in the unhappy position of being a strong supporter of passive obedience just at the time when James II's policies were provoking the political nation to rebellion.[36]

Reformist schemes aroused the hostility of those who stood in the way of the arbitrary efficiency to which they aspired, either the watchdogs of constitutional privilege or the protagonists of sectional interests. When John Collins devised an impressive scheme to raise revenue from a duty on salt to support the fishing industry and other public-spirited enterprises, he felt obliged to incorporate safeguards to convince Parliament that it would not give a dangerous and unchecked boost to royal finance, which Parliamentarians rightly saw as the key to independent and arbitrary executive power. There are also scattered hints of alarm at the possible threat of technological schemes to vested interests. In his vociferous onslaught on the Royal Society, Henry Stubbe claimed that a proposal had been made to Parliament 'that such kinde of *pretended new Inventions* relating to *Mechanicks*, *Trades*, or *Manufacturers*, as *are* or *shall be* offered to the *Parliament*' should be adjudicated by the council of the Royal Society. He suggested that the Society would soon be wishing to '*detect the Frauds of Trades*', and he darkly warned the tradesmen of London to consider the consequences of such proposals.[37] A Huntingdonshire yeoman voiced comparable reservations about the Society's concern with agricultural improvement, refusing to reply to its inquiries on the grounds that 'there is more reason of state in this Royall societie then at first I was aware of'. A like suspicion is implicit in the mistake of those who took Edward Lhwyd, on his natural historical perambulations around Wales, if not for a Jacobite spy, then for a tax-collector.[38]

---

[36] Pett, *The Happy Future State of England* (London, 1688), pp. 112, 64 and passim; Pett, *The Obligation Resulting from the Oath of Supremacy* (London, 1688); Jacob,[221] pp. 34–5.   [37] Cl. P. xxv, fol. 115; Stubbe, *P.U.*, sigs. b2v–3.

[38] Tenison to Oldenburg, 7 Nov. 1671, Oldenburg,[121] VIII, 348; Lhwyd to John Lloyd, 28 Feb. 1698, Gunther,[130] p. 355.

Similarly, Aubrey was aware of likely hostility to an enlightened educational establishment, which others who tried to set up such schemes actually encountered.[39] It is not surprising that Lewis Maidwell, a projector who hoped to finance such a plan with a publishing monopoly, was dealt with particularly severely, in view of the jealousy with which the Stationers' Company protected its monopoly: he was, however, evidently seriously considered by the Crown. The Royal Society had earlier considered an analogous proposal that its revenue might be provided by a levy on all scientific books published in England, which might have raised similar hackles had it been taken any further.[40]

This is also the background to the antagonism of the established corporations to these ambitions, and hence to the Royal Society which exemplified them. Few features of Restoration political history are more notable than the Crown's attack on towns, colleges and other institutions as bastions of resistance to royal power. Hatred of the Society's pretensions at Oxford clearly owed something to this, and, in view of the King's use of his dispensing power to fill university posts, Stubbe's hint that the Royal Society might want to recommend people for academic preferment was most damaging.[41] Still clearer is the case of the London College of Physicians, who felt particularly vulnerable in view of the rather precarious monopolistic position of its licenciates among the medical practitioners of the city: thus Stubbe accused the Royal Society of fomenting the quarrel between non-licenciates and the College. Such bodies feared that reforms espoused by the Crown might infringe their valuable rights and privileges. Significantly, the academic and medical professions were the very groups most threatened by scientifically-oriented reform in the Interregnum, and both were the object of government interference in the reign of James II. The College of Physicians had its Fellowship purged and enlarged when its charter was regranted (quite apart from suffering an embarrassing inquiry into its claims to privilege at the end of the reign), while at Oxford the King's confrontation with the Fellows

---

[39] Aubrey 10, esp. fol. 146; Caudill,[147] ch. 8, passim.
[40] Caudill,[147] p. 342; Birch,[118] I, 377.
[41] Western,[6] pp. 69–77; J. H. Sacret, 'The Restoration government and municipal corporations', *English historical review*, 45 (1930), 232–59; E. F. Churchill, 'The dispensing power of the crown in ecclesiastical affairs', *Law Quarterly review*, 38 (1922), 309f; Stubbe, *P.U.*, sig. b3.

of Magdalen College was one of the *causes célèbres* of the 1688 Revolution.[42]

It is even possible that it was because of the threat that science seemed to pose to vested interests that lawyers were proportionately rare in the ranks of the Royal Society, though those who belonged made up for this by being disproportionately active. It was a commonplace that 'so few Gentlemen of the Law were noted for loyalty . . . that it was made a Wonder at Court that a young lawyer should be so'.[43] John Beale thought that Bacon's rationalising attitude to the law – an integral part of his reformist programme for intellectual life – still rankled in legal circles, while Aubrey, for one, would have liked to see the law reorganised.[44] Arguments for such antipathy are, however, speculative, not least since an attempt to systematise the law along inductive lines – itself possibly influenced by science – originated within Restoration legal circles.[45]

What is undoubtedly revealing, in view of their appeal to authority and their admiration for effective government in France, is the stance of men devoted to science and economic improvement in the Exclusion Crisis (1678–81). For during this open challenge to the authority of the Crown and the inviolability of the succession by the nascent Whig party, John Houghton and others supported the Tories and the court.[46] Furthermore it was in the 1680s, when England came closer to absolutism than at any other time in this period in the reaction that followed the failure of the Whigs, that natural philosophy came nearest to achieving the 'established' position that English scientists had desired ever since the Restoration. The setting was a general movement towards centralisation and efficiency seen in the reform of the armed forces and a streamlining of government administration which was linked to the improved financial position enjoyed by the Crown.[47]

Most striking, in 1682 Charles II bought back from the Royal Society the rights of ownership to Chelsea College, which he had

---

[42] Clark,[251] chs. 17–19; Brown,[253] pp. 16–20; Stubbe, *Camp.*, p. 20; Webster,[72] chs. 3–4; Western,[6] pp. 200–2; Jones,[222] pp. 119–21.

[43] Hunter,[107] p. 38 and tables 5–6; Michael Landon, *The triumph of the lawyers 1678–89* (Alabama, 1970), p. 39; Western,[6] p. 55.

[44] P. H. Kocher, 'Francis Bacon on the science of jurisprudence', *J.H.I.*, 18 (1957), 3–26; Beale to Evelyn, 9 May 1668, Evelyn, *Corr.*, no. 73; Hunter,[143] pp. 52–3.

[45] B. J. Shapiro, 'Law and science in seventeenth-century England', *Stanford law review*, 21 (1969), 727–66.   [46] Jacob,[221] pp. 30–2.

[47] Western,[6] chs. 4–5; Tanner,[184] p. 73; Tomlinson in Jones,[2] pp. 95, 100–1; Chandaman,[224] pp. 251–61.

granted to the Society in the 1660s but which had since proved more a liability than an asset due to protracted disputes over gaining possession. Charles now needed the site for the military hospital which he established on the model of Louis XIV's *Invalides*, which still survives. In return, he paid the Society £1,300, which was invested in East India Company stock (in itself perhaps not an insignificant gesture, since courtly favour to this company had been attacked by the Whigs as a symptom of royal corruption during the Exclusion Crisis).[48] Thus the Society gained the regular income, other than that from Fellows' subscriptions, which it had always wanted, though (as we saw in chapter 2) by this time it had settled down to an institutional life of an unimpressive kind to which the new endowment made little difference.

But it is interesting that in the reign of James II there were hopes that the Society might at last gain all it wanted. Early in 1686 it seemed likely that the King might grant the Society a piece of land in St James's Park on which to build its long-desired college, though in fact nothing came of this. Later in the same year it appeared possible that the newly-formed ecclesiastical commission would reallocate the endowments of Gresham College (a typical authoritarian gesture) and that the Royal Society might benefit from this, though in the event this, too, proved a disappointment.[49] No less significant are Petty's fortunes under James II. For Petty gained a new influence at court in 1686 and seems genuinely to have believed that it might now be feasible to implement some of the proposals for efficient economic planning that he had long suggested. Petty's premature death cut short such hopes, but the episode stimulated him to formulate several of his reformist schemes.[50]

The fortunes of science at Oxford in these years were equally remarkable. The Ashmolean Museum, symbol of the University's official espousal of the new philosophy, was formally opened by James as Duke of York in May 1683 – in the very summer that Convocation proscribed the doctrine of mixed monarchy and various related principles and ordered tutors to preach passive obedience, and that the City of Oxford surrendered its municipal charter and signed an instrument giving the King power over

[48] Bluhm,[108] p. 85; Jacob,[221] pp. 29–30. See above, p. 40.
[49] Evelyn to Pepys, 1 Mar. 1686, 28 Nov. 1686, Evelyn, *Corr.* no. 1549, Evelyn, *Lb.*, no. 552.
[50] Strauss,[217] pp. 131–3; Miller,[223] pp. 167–8; Hull,[233] II, 631n.

elections.[51] Robert Plot, first Curator of the Ashmolean, was himself a Catholic, patronised by various of James II's favourites, who fraternised with the King when Oxford loyally welcomed him in September 1687 and was appointed Historiographer Royal in 1688. Plot even pandered to the King's absolutist proclivity in a blueprint for scientific research in a utopian setting, promising an inexhaustible supply of gold from alchemical transmutations which would make parliaments permanently unnecessary.[52] Of course, not all went so far, and most scientists, like others in the country, were repelled by the increasing extremity of James's religious policy towards the end of his reign: but this juxtaposition of science and the movement towards absolutism is striking.

Though it is significant that science reached this high point of official patronage in the reactionary 1680s, however, it is more revealing that the government's record in sponsoring science and in taking up the optimistic plans of scientific reformers fell far short of their hopes. There were various reasons for this. In part it was due to a hard-headed suspicion of the plans of 'projectors' in government as in economic life, perhaps encouraged by Petty's own participation in rather questionable schemes. It also owed much to political realism, for the experience of the Interregnum had left a deep suspicion of change among the Members of Parliament on whom royal revenue depended for much of the period. Indeed, Petty's idealism was a measure of his lack of political acumen: 'I cannot dance upon the Roapes of criticall honor and Courtshipp', he complained to his friend Sir Robert Southwell, and he wrote a hostile account of 'the Art of a Courtier', who made the chief governor absolute, but without doing him or the country any service.[53] Most of all, however, it was due to the character of Restoration government, which reflected the inclination of Charles II and his ministers – and, to a lesser extent, his successors and theirs – to do what was easiest, so that reforms were passed by and more piecemeal expedients

[51] C. E. Mallet, *A history of the University of Oxford* (London, 1924), II, 448; J. P. Kenyon, *The Stuart constitution* (Cambridge, 1966), pp. 471–4; Wood, *Life and Times*, III, 52, 55, 62–4, 67, 75–8, 86, 89, 112.
[52] Anthony Wood, *Athenae Oxonienses*, ed. P. Bliss, IV (London, 1820), 773–4; Mallet, *A history . . .*, II, 452; Wood, *Life and Times*, III, 232; Gunther,[129] p. 412.
[53] Colin Brooks, 'Taxation, finance and public opinion 1688–1714' (Cambridge Ph.D. thesis, 1971), p. 7; Strauss,[217] pp. 125–7, 142–3 and chs. 10–11 passim; Petty to Southwell, 4 Jan. 1679, Lansdowne,[235] p. 62; Lansdowne,[234] II, 189–90.

preferred. Though it is possible to caricature governmental administration under the later Stuarts, there is no doubt that it fell far short
of the high ideals that Petty and others set. Despite the growth of a
business-like executive and an improvement of standards, particularly in the armed forces, much inefficiency remained; still more
pernicious was the dominance of amateurs over professionals, which
stunted the adoption of long-term policy.[54] Hence governmental
patronage of science was apathetic, as Petty and other would-be
reformers discovered to their cost.

This was clearest under Charles II, with muted echoes in the
subsequent reigns. Charles was a classic virtuoso. He had imbibed a
fashionable curiosity about the new science during his French exile
and 'knew of many Empyrical Medicines, & the easier Mechanical
Mathematics'. He had a laboratory in his palace at Whitehall where
experiments were carried out by a royal chemist and his staff and he
could boast a cabinet of curiosities similar to those to be found
throughout London: Evelyn was called on for advice in cataloguing
this in 1665.[55] In 1661 the King held Petty '½ an hower before 40
lords upon the philosophy of Shipping, loadstone, skreen'd [?] guns,
the feathering of arrowes, the vegetation of plants, the history of
trades etc.' and, perhaps encouraged by the King's example, science
also flourished at court. In 1665 evenings were passed with lectures
by Petty and others 'upon something that nobody understands but
themselves', in one courtier's words.[56] Scientific papers were occasionally read and curiosities displayed at court later in the reign, while
in 1677 Robert Hooke was summoned there to give a demonstration of
the 'little animals' that Antoni van Leeuwenhoek had seen through
his microscope in pepper water. Many courtiers had scientific interests like the King's, as also did his cousin, Prince Rupert.[57]

The King and his courtiers shared the conviction of other virtuosi
as to the utility of scientific pursuits. The King solemnly told Petty

[54] Western,[6] pp. 101–2, 125–9; Tanner,[184] pp. 27, 77; Tomlinson in Jones,[2]
pp. 99–100, 102, 116; Chandaman,[224] ch. 6.
[55] Evelyn,[164] IV, 409, III, 300; Pepys,[163] IX, 415–16; Samuel Sorbière, *Relation d'un
Voyage en Angleterre* (Paris, 1664), pp. 80–1, 84–6, 97; Balthazar de Monconys,
Iournal des Voyages (Lyons, 1666), II, 39, 57–8, 77; Evelyn to Chiffing, 4 Feb.
1665, Evelyn, *Lb.*, no. 233. See also Sprat,[92] p. 133.
[56] Sharp,[144] p. 169; Hunter,[107] p. 14.
[57] E.g. Deacon,[182] p. 127; Ray to Robinson, 9 Apr. 1685, LBO x, fol. 89; Hooke,[165]
pp. 335–6 (cf. ibid, p. 338); Hartley,[97] pp. 147–57, 159–65, 199–210, 239–50; A.
Robertson, *The life of Sir Robert Moray* (London, 1922), ch. 8 and pp. 181f; K.
Dewhurst, 'Prince Rupert as a scientist', *B.J.H.S.*, 1 (1963), 365–73.

and Hooke to apply themselves to shipping, showing an awareness of the potential of new inventions and discoveries for naval affairs that impressed Pepys.[58] Similarly, Prince Rupert devised a number of improvements in weaponry, some of them thought highly of by contemporaries, while Sir Joseph Williamson, Secretary of State, told Hooke in 1677 'to be diligent for this year to study things of use'.[59] It would be gratifying if it could be proved that this was intensified when Charles's more serious brother came to the throne, but there is no evidence of this, though James had taken an interest in naval inventions when he was Lord High Admiral while Duke of York.[60]

But there was a certain ambivalence in Charles's concern for science. Despite his philosophical dabblings, he could not resist poking fun at his scientific subjects. Pepys recorded in 1664 how 'the King came and stayed an hour or two, laughing at Sir W[illiam] Petty, who was there about his boat, and at Gresham College in general . . . for spending time only in weighing of ayre, and doing nothing else since they sat'. In 1668 gossip in the Low Countries (recounted with surprise by an Italian visitor) told that the King was accustomed to refer to the virtuosi of the Royal Society as his 'fous', his jesters.[61] His attitude was echoed by the ridicule of the 'court wits' which annoyed those devoted to science. 'If a Grandee scoffes, the followers must not be too large at the Panegyric', John Beale thought, and his letters show constant concern about the effect of satire on likely supporters, 'since Political persons must have some regard to public Fame'.[62] Scientific interests in these circles were apparently not very sustained: they tended to be recreational and of no great relevance to government. Though attention has often been drawn to the curiosity of civil servants like Pepys, William Blathwayt and William Lowndes about science and its potential for government, too much is easily made of this since such men were in a minority and their influence limited.[63]

[58] Sharp,[144] p. 169; Hooke,[165] p. 185; Pepys,[236] p. 115; cf. Hales, *Account*, pp. xxf.
[59] Dewhurst, 'Prince Rupert', pp. 367–9, 371–2; Hall,[173] pp. 11, 61; Hooke,[64] p. 22; Defoe,[202] pp. 25–6; Rawlinson A 178, fol. 59; Hooke,[165] p. 337.
[60] D. McKie, 'James, Duke of York, F.R.S.', *N.&R.R.S.*, 13 (1958), 7; Monconys, *Iournal des Voyages*, II, 40–2. See also above, p. 106.
[61] Pepys,[163] V, 32–3; W. E. K. Middleton, 'What did Charles II call the Fellows of the Royal Society?', *N.&R.R.S.*, 32 (1977), 14.
[62] Beale to Evelyn, 1 Nov. 1671, Oct. 1669, Evelyn, *Corr.*, nos. 123, 89.
[63] Plumb,[5] pp. 11–12; Jacobsen,[225] pp. 103–4; Buck,[219] pp. 80–1; Tomlinson in Jones,[2] pp. 115–16; Jones,[222] p. 28.

In these circumstances it is hardly surprising that the Royal Society's pleas about what it might have achieved with proper finance fell on largely deaf ears: parsimony thwarted scientific co-operation, as when the Navy Commissioners suggested that the Society should help with weighing up wrecks in the Thames and received the reply that it was 'destitute of the necessaries for such a work'. The same is true of the reformist aspirations of men like Petty, for the King and the government had little time for his ambitious schemes. Charles complained how Petty was always 'Ayming at Impossible things', while court gossip accused him of being 'notionall and fancy-full neer up to madness', if not 'a Conjurer' or 'a Fanatick'.[64] When Petty suggested a remodelling of the taxation system that would have greatly benefited the government in the long term, his suggestion was rejected in favour of existing arrangements. Suspicion, incompetence and niggardliness prevented him from getting tax officials to gather the accurate information that he wanted even in Ireland.[65] So too, because of the jobbery and place-seeking by which offices were allocated and despite the obvious importance of *Sylva*, Evelyn failed to obtain an official inspectorship of timber that he desired: instead it 'was conferred upon another, who, I believe, had seldom been out of the smoke of London', as Evelyn bitterly reflected.[66]

The government occasionally patronised science and scientifically-inspired reform. We hear of pensions granted to two students of applied mathematics: Henry Bond was given £50 a year for life and a young Oxford graduate was sent to the Netherlands to study fortification techniques. The King also encouraged Halley's scientific voyage to the South Atlantic in 1676–8, partly by persuading the East India Company to assist him, though the outlay involved was very limited by later standards.[67] Rather more revealing, however, are two other events of Charles's reign, the inauguration of the Royal Observatory in 1675 and the foundation in 1673 of the Royal Mathematical School, which was attached to Christ's Hospital with the intention of training seamen for the navy.

The initiative for both the school and the observatory came from individual civil servants, Samuel Pepys and Sir Jonas Moore. In the

[64] Birch,[118] II, 385; Southwell to Petty, 13 Aug. 1687, Petty to Southwell, 16 Sept. 1682, Lansdowne,[235] pp. 281, 103–4.

[65] Sharp,[144] pp. 350–1; Strauss,[217] p. 130; Lansdowne,[234] I, 103–11.

[66] Evelyn to the Countess of Sunderland, 4 Aug. 1690, Evelyn,[126] p. 318.

[67] W. A. Shaw, ed., *Calendar of Treasury Books 1676–9* (London, 1911), p. 287; Collins to Bernard, 16 Mar. 1671, Smith 45, fol. 61; Halley,[238] pp. 2–3, 179–81.

case of the Royal Mathematical School the government only reluc-
tantly provided an endowment and its indifference enhanced the
school's severe problems in its early years, which greatly disappointed Pepys, who saw its potential value to the navy. Other
problems, however, included a lack of enthusiasm among the merchant governors and the difficulty of finding masters with a suitable
combination of intellectual and practical ability (for academics
tended to use the post as a sinecure while sea-tars lacked educational
background).[68] The observatory's history is comparable, for, though
this was ultimately of great importance, its early years show a
half-heartedness on the part of the government that is almost as
striking as the original foundation. Flamsteed complained bitterly of
the government's parsimony in providing instruments and in 1680
was in fear of 'total retrenchment' of his allowance. He had to
augment his salary by taking pupils in the early years, and after his
father's death in 1688 he supplemented both his salary and his
equipment from his private means.[69]

Changes in official attitudes might have been expected in the
1690s, when progress towards rationalisation and efficiency was
accelerated by the unprecedentedly heavy financial and organisa-
tional burden imposed by William III's European wars. It is true that
the political arithmetic that Petty had pioneered now came to play a
part in government for the first time as statistics – supplied, among
others, by John Houghton – became vital instead of mere scholarly
curiosities. An Inspector-General of Exports and Imports was
appointed in 1696, and the statistical records that he kept still
survive, while the government also sponsored Edmond Halley in
compiling efficient life-tables, which were needed to calculate the
annuities on which war loans were raised.[70] Halley also presided in
1698 over 'the first purely scientific maritime expedition dispatched
by the English government' – a royal ship was provided so that he
could investigate magnetic variation in the Atlantic – while intellec-

---

[68] Taylor,[156] pp. 114–19; Pearce,[230] ch. 6; Allen,[231] p. 44 and ch. 2 passim; Arthur
Bryant, *Samuel Pepys: the years of peril* (Cambridge, 1935), pp. 363–5, 375–6,
386–8; Pepys,[236] pp. 102, 136, 148–9.

[69] Flamsteed,[162] pp. xxvi–xxx, 45n, 118–19 (Flamsteed to Ward, 3 Jan. 1680);
Forbes,[183] pp. 29, 38; Weld,[94] I, 255–6.

[70] Charles Davenant, *Discourses on the Public Revenues* (London, 1698), I, 1–35;
G. S. Holmes, 'Gregory King and the social structure of pre-industrial England',
T.R.H.S., 5th series 27 (1977), 45 and n.20; Clark,[228] pp. xiii, 1–4; Clark,[172]
pp. 136–7, 138–9.

tuals were deliberately enrolled to a Board of Trade set up in 1696,
evidently in the hope that they could transcend party interest in
finding improvements on which all could agree.[71]

But it is easy to overestimate the significance of these developments even then. Paradoxically, the statistics most commonly cited
for this period by modern scholars were compiled by Gregory King,
who was hostile to the war and its impact. Even such changes as the
war produced were not all permanent and one modern author has
noted a slackening in the use of quantitative methods and a 'general
administrative torpor' in the eighteenth century.[72] The Revolution
of 1688 has often been interpreted as a victory for conservatism,
representing a defeat for impulses towards efficient bureaucracy in
favour of the preservation of anomalies exploited by sectional interests and the continued dominance of amateurs in government.
Even the Board of Trade of 1696 performed disappointingly, due
largely to inter-departmental tensions.[73]

So, on the whole, official patronage for scientists and their
schemes in the Restoration was disappointing, and their far-reaching
but politically dangerous aspirations to reform never came to fruition. As with technology, it was difficult for intellectuals to play the
part they would have liked, and for those in government scientific
interests remained essentially separate from 'secular affaires, which
is the *Burial* of all *Philosophicall Speculations* & Improvements', in
John Evelyn's words.[74] This evidently explains why there are hints of
an alternative to the ambitious strategy of Petty in scientific circles,
which advocated science's deliberate insulation from public affairs.
It was, perhaps, the administrative equivalent of the retreat from
technology noted in the last chapter, with indolent enthusiasts
preferring to see science as safe, neutral and (if necessary) insignificant. It was thus claimed that it was not the Royal Society's business
to interfere in 'state concern[s]', even if this meant eschewing vital
matters like navigation. A similar mentality underlay the Society's
reply when asked to support a projected trading company in the
1690s (an idea that would have appealed to Petty), that 'this being a

[71] G. M. Rodgers, 'Official English maritime explorations 1660–1780' (Texas Ph.D.
thesis, 1958), p. 60 and pp. 61–70 passim; Halley,[238] pp. 8, 103f, 243–7; Wilson,[13]
p. 167; Laslett,[227] esp. pp. 139–40.
[72] G. S. Holmes, 'Gregory King', pp. 51–2; Clark,[172] pp. 144–5.
[73] Jones,[222] p. 16; Holmes,[3] p. 10; Brooks, 'Taxation, finance and public opinion', chs.
3, 5 and passim; I. K. Steele, *Politics of colonial policy* (Oxford, 1968), esp. ch. 2.
[74] Evelyn to Beale, 11 July 1679, Evelyn, *Lb.*, no. 406.

matter concerning trade itt was thought not proper for the Society to medle in itt'.[75]

Neither such withdrawal nor the disappointing outcome of scientists' schemes, however, reduces the significance of the aspirations of men like Petty. There can be no doubt about science's political flavour in its support for centralised power and its opposition to sectional interest. Science was, of course, a marginal issue in the age of the Exclusion Crisis and the Glorious Revolution, and it is wrong to force its affiliations too neatly into party labels: if scientists veered in a Tory direction during the Exclusion Crisis, they moved towards the Whigs in the 1690s. Their loyalty was always to those in government for the simple reason that there things were achieved: perhaps the most relevant political division is that between 'court' and 'country', in which backbench Whigs and Tories often joined in their opposition to the executive.[76] Science and strong government – in England as in Europe – had a natural affinity. But their co-operation was thwarted by the constitutional opposition and administrative pragmatism that define the true character of Restoration politics.

[75] Birch,[118] 1, 249; JBO x, fol. 38.
[76] H. T. Dickinson, *Liberty and property: political ideology in eighteenth-century England* (London, 1977), ch. 3.

6

Science, learning and the universities

For various reasons, as shown in the last two chapters, the ardent desire of Restoration scientists that natural philosophy should play a wider role in the running of the country was disappointed. Though its wider aspirations may have affected its reputation, science was always primarily a theoretical activity and its most significant contemporary milieu therefore in intellectual life. This chapter will survey the relations between the new science and other learned traditions and their institutional background, assessing the friction that arose, its continuity and its implications.

Opinions on this have differed. Some have seen the universities and the spokesmen of the scholarly traditions that they stood for as irrevocably opposed to the new science; others have argued the contrary. Evidence presented in chapter 3 of this book has already hinted at the latter view: most major scientists had an academic background and had often had the opportunity to develop their scientific interests while at Oxford or Cambridge; a number pursued academic careers, while the prominence of doctors among scientific researchers suggests harmony with medicine. On the other hand, however, the new science encountered lack of encouragement and even opposition in academic and medical circles, which caused its supporters concern and which has often attracted comment since. It has become a historiographical commonplace that the Royal Society and institutions like it in other countries gave organised support to the new philosophy which was lacking in traditional academic settings.[1] It is equally familiar that the new science faced much antagonism: apart from a series of published attacks, hints of public and private statements suggest more widespread hostility.

The two most prominent figures in this quarrel were Henry Stubbe

[1] E.g. Ornstein,[98] pp. 257–63; Purver,[24] ch. 3.

and Meric Casaubon. Stubbe, a Warwick doctor who had already established himself as a prolific pamphleteer, published a string of tracts with sensationalist titles against the Royal Society in 1670 and 1671. Casaubon was a prebendary of Canterbury and a distinguished scholar and divine, son of the famous classical scholar, Isaac Casaubon. He brought out a single, more sober pamphlet on this topic, though he echoed the ideas expressed there in other books written at this time and in an unpublished tract dedicated to Francis Turner, Bishop of Ely; this has only recently been discovered and has not received the attention it deserves.[2] Though neither Stubbe nor Casaubon was an academic, both clearly felt that they were defending the universities and the learned traditions of the time, though they also put forward wider arguments. 'God preserve the Universities', Casaubon exclaimed, and it may be significant that his attack on science was printed at Cambridge.[3] Stubbe dedicated one of his tracts to the universities and he enjoyed considerable support at Oxford. Most spectacularly, he was spontaneously given a piece of plate worth £5 or £6 by Thomas Pierce, President of Magdalen College, and his onslaught was said to be approved by the influential John Fell, Dean of Christ Church, to whom another of his pamphlets was dedicated.[4]

Another challenge to the Royal Society and the new science came from Robert South, Public Orator at Oxford, in a speech at the dedication of the Sheldonian Theatre there in 1669. South echoed this in various sermons and his speech at Oxford was paralleled by similar ones at Cambridge which drew large audiences. Also at Cambridge, Peter Gunning, Bishop of Chichester and Regius Professor of Divinity, refused to license a work which included verses in praise of the Society in 1669, while in 1672 a book printed at Oxford had to have words 'somewhat in favour of the Royall Society, & of new Philosophy' 'a little mollified'.[5] Another influential Oxford figure, Thomas Barlow, Bishop of Lincoln, apparently took excep-

---

[2] Rawlinson D.36.1. On its discovery and identification see Spiller,[243] esp. pp. 141–5.

[3] Casaubon, *Of Credulity and Incredulity, In things Natural, Civil, and Divine* (London, 1668), p. 300.

[4] Stubbe, *Legends*, sigs.† 4, A1–2; Wallis to Oldenburg, 28 Oct. 1670, Oldenburg,[121] VII, 225; Stubbe, *Censure*, sig. [a]2; Wood, *Life and times*, I, 354.

[5] Wallis to Oldenburg, 16 July 1669, Oldenburg,[121] VI, 129–30 (see also ibid, p. xxvi); Wallis to Boyle, 17 July 1669, Boyle,[71] VI, 459; Syfret,[239a] pp. 240f; Stubbe to Boyle, 4 June 1670, Boyle,[71] I, xcv; du Moulin to Boyle, 3 July 1671, ibid, VI, 580; Wallis to Oldenburg, 18 Jan. 1672, Oldenburg,[121] VIII, 483.

tion to scientific ideas, while Obadiah Walker, Master of University College, expressed reservations about the claims of the new science to novelty and importance.[6]

There was also opposition to the new science in medical circles. Stubbe, a doctor himself, claimed that the Royal Society's threat to his profession contributed to his decision to launch a public assault, in which he said that many physicians encouraged him. A story, however, that he was hired by Baldwin Hamey, President of the College of Physicians, to attack the Society is recorded only in a late source of dubious reliability (the same applies to an anecdote that Hamey tried to interest the Society in a spurious gadget as a joke).[7] Antipathy towards the new science is certainly illustrated by Thomas Wharton, Censor of the College, who fulminated against 'this new upturned brood of Vertuosis' with their 'Jesuitisme & policy, English-books, Experiments, & receipts in phisick', who were likely to influence 'all the families of Note in England'; he linked them with 'the swarmes of quackes, mountebacks, chymists, Apothecares, surgions' who were ruinous to 'our old and settled and approved practice of physick'. Wharton was also hostile to the challenge to the 'old & sound' Aristotelian learning of the Universities of the 'new fangled fopperies' of Descartes, Gassendi, Boyle, Hobbes and Regius.[8]

These articulate attacks on science voiced or aroused hostility to the new philosophy and the Royal Society beyond the ranks of academics and physicians. Here other accusations entered, notably that of atheism, which will be separately considered in the next chapter, but the threat to the intellectual traditions of the day that the new science was claimed to present was not irrelevant. It was thus claimed that Stubbe's abuse made the common people of Bristol believe that the Society comprised 'a Company of Atheists, Papists Dunces & utter enemies to all learning'.[9]

---

[6] Thomas Barlow, *The Genuine Remains* (London, 1693), pp. 151–9 (for earlier manuscript versions of the first of these letters attributing it to one 'L.M.', evidently an Anglo-Irish cleric, see Sharp,[144] pp. 318–19, 412: but since the second refers to the first and is clearly by Barlow I do not know what to make of this); Walker,[268] pp. 118–19.

[7] Stubbe, *Camp.*, p. 8; Stubbe to Boyle, 17 Dec. 1669, 4 June 1670, Boyle,[71] I, xci, xcv; Glanvill to Oldenburg, 19 July 1669, Oldenburg,[121] VI, 138; John Keevil, *The stranger's son* (London, 1953), pp. 178–9.

[8] Wharton to Mrs Church, 15 May 1673, Royal College of Physicians of London MS 640, fol. 10.   [9] Glanvill to Oldenburg, 31 Jan. 1670, Oldenburg,[121] VI, 456.

Henry Oldenburg thought that Stubbe 'must be soundly lashed to undeceive the Vulgar', and at Stubbe's death in 1676, Thomas Mariet, a Warwickshire lawyer, commented: 'He is by many Lamented, & by some as holding a rodd ouer the Royal Society whoe for him durst not trouble the world with their impertinencyes.' Mariet's surviving letters show that he himself felt some scepticism about science.[10] Indeed Stubbe declared that 'most sober persons' commended his design and that he had drawn the attention of former supporters of the new philosophy to its pernicious implications: this is plausible in explaining the vindications that Glanvill and others addressed to those 'worthy Persons' whom they claimed Stubbe had misled.[11] John Beale was especially worried about Casaubon and the 'huge Clatter' made by unfavourable remarks from 'so great a Name amongst the Critics & Antiquaries'. A general dislike of the new science is also indicated by Sprat's and Glanvill's belief that their design needed justification to all classes against 'Objections and Cavils' from the 'many *Criticks*' 'of whom the World is now full'.[12]

It is essential, therefore, to assess what anxieties underlay this opposition, how real they were and how widespread, in the context of science's place in the intellectual scene. It is important to allow for complexity in such attitudes and to assess the reliability of science's adversaries. Casaubon was a churchman, as were other prominent antagonists, and his concern was particularly with religion. But it would be wrong to assume that such matters caused anxiety only to ecclesiastics, as is shown by the evidence just cited of disquiet among the laity, and Casaubon argued that his scruples had implications for secular affairs.[13]

Stubbe is more problematic. He was a ready pamphleteer, quick to make cheap points at his victims' expense, who wrote polemical works on numerous subjects during his career: one spokesman for the Royal Society threatened him with a survey of these to 'shew in how many press-wars you have served as a volunteer'.[14] It has

[10] Oldenburg to Evelyn, 8 July 1670, Oldenburg,[121] vii, 58; Mariet to Aubrey, 14 Nov. 1676, Aubrey 12, fol. 327; Hunter,[143] p. 135.

[11] Stubbe to Boyle, 18 May 1670, Boyle,[71] i, xcii; *A Brief Vindication of the Royal Society* (London, 1670), p. 1. See also Glanvill, *A Praefatory Answer to Mr Henry Stubbe* (London, 1671), *A Further Discovery of M. Stubbe* (London, 1671).

[12] Beale to Evelyn, 10 May 1669, Evelyn, *Corr.*, no. 82; Sprat,[92] sig. B4v, p. 46.

[13] Casaubon, *Of Credulity and Incredulity*, p. 296.

[14] *A Letter to Mr. Henry Stubs* (London, 1670), p. 19.

recently been argued that, although Stubbe claimed to defend orthodox values against the virtuosi, he was really a deeply subversive
figure who combined heretical philosophical views with republican
principles: though only suggestions of these are to be found in his
published writings, they can be supplemented from unpublished
works.[15] But the hints of this kind in his attack on the Society are so
slight that they eluded both contemporaries and the majority of
commentators. To most Stubbe seemed – whatever his ulterior
motives – to be articulating commonplace reservations about the
new science which were widely held if rarely so fully expressed.
Indeed, in his anxiety to gain support he arguably made these views
more than usually explicit. Moreover Stubbe's polemics apparently
stemmed from a more local dispute in the West Country between
Glanvill and Robert Crosse, an Aristotelian, which Stubbe had not
begun.[16]

It is also crucial not to assume a simple polarisation between a
uniform party supporting the new science and the Royal Society and
an academic orthodoxy hostile to both. Opinions differed in scientific circles on the relations between science and other forms of learning, as on so many other topics, and some were undoubtedly more
extreme than others. Casaubon believed that all in the Society would
not stand by the claims of certain propagandists, while Stubbe
claimed to attack 'these *Prattle-boxes*', the lesser virtuosi, rather
than serious scientists like Boyle, Hooke and Wren, though (as we
shall see) it is doubtful if so convenient a distinction is feasible.[17] On
the other hand, there were clearly differing estimates of science at the
universities, as is illustrated by Anthony Wood's comment when the
Doctors and Masters viewed the contents of the newly-opened
Ashmolean Museum at Oxford in 1683: 'Many that are delighted
with new philosophy are taken with them; but some, for the old,
look upon them as baubles.'[18] These complications must be borne in
mind in assessing how fundamental and permanent any hostility
may have been.

---

[15] J. R. Jacob, 'Civil religion and radical politics: Stubbe to Blount' (unpublished
paper). See also Zagorin[79], pp. 159–62; P. M. Holt, *A seventeenth-century defender
of Islam: Henry Stubbe (1632–76) and his book* (London, 1972).
[16] Anthony Wood, *Athenae Oxonienses*, ed. P. Bliss, IV (London, 1820), pp. 123–4;
Jones,[242] pp. 254–5. See also above, p. 8.
[17] Casaubon,[265] p. 3; Stubbe, *P.U.*, sig. a3, p. 166; *Legends*, sigs. *4, A3v. See below,
p. 151.
[18] Wood, *Life and times*, III, 56.

One thing is certain, that the Restoration universities were not monolithically opposed to science. Oxford's central role in Interregnum science has already been indicated, and research begun then continued after 1660.[19] Furthermore, scientific studies were widespread in the Restoration, though they penetrated undergraduate study only haphazardly. Science was pursued mainly among graduates and more at an unofficial than an official level, a fact that has led to much misunderstanding among modern scholars. But plenty of scientific facilities were available, as was pointed out to critics of the university for its neglect of the new philosophy. For much of the 1660s Oxford was the chosen home of Boyle, who was said to be 'very communicative' to those interested in chemistry and other experimental pursuits, while chemistry classes continued after the Restoration as before. John Wallis taught mathematics throughout the period, while lectures on mathematics and astronomy survive that Edward Bernard and David Gregory delivered as Savilian Professors of Astronomy. At the Botanical Gardens, founded in 1622, Robert Morison and others gave instruction in botany, and anatomy lessons were offered by Richard Lower, Thomas Willis and William Musgrave.[20] Now, as in the Interregnum, the collegiate system was a focus for scientific as other activities, and some colleges could offer more than others: thus Trinity provided scientific and mathematical instruction even to undergraduates under the aegis of its scientifically-inclined President, Ralph Bathurst, and there is similar evidence from other colleges, some of which also had their own virtuoso cabinets.[21]

At Cambridge, evidence is hardly less convincing. Indeed soon after the Restoration the new philosophy was said to be 'much in all people's discourse', more so than at Oxford. The Cartesian teaching begun by Henry More in the Interregnum became more widespread in the Restoration, while a textbook on the subject was published there in 1686.[22] Mathematics was also well established: the Lucasian

---

[19] Frank,[40] esp. chs. 6–7; R. G. Frank, jr, 'The John Ward diaries: mirror of seventeenth-century science and medicine', *Journal of the History of Medicine*, 29 (1974), 147–79.

[20] Wallis,[266] pp. 315–17, 320–3; Frank,[132] pp. 200–4 and passim; Shapiro,[250] pp. 67–75.

[21] H. E. D. Blakiston, *Trinity College* (London, 1898), p. 173; Frank,[75] pp. 194–5; Frank,[132] pp. 256–61; Caudill,[147] pp. 353–4; Allen,[247] pp. 129–30.

[22] H. C. Foxcroft, *A supplement to Burnet's History of my own time* (Oxford, 1902), pp. 46–7; Kearney,[244] p. 151.

professorship held by Newton was founded in 1663, and the percolation of such interests through the university is indicated by the catalogue of a young don's books in the 1670s, which includes twenty-three mathematical titles out of a total of 120.[23] By the 1680s a chemist, John Vigani, was lecturing at Cambridge, and there are hints of 'a Club of Philosophicall Chymists' earlier. Again, some colleges were probably better off than others for scientific facilities: Trinity could boast a chemical laboratory and a set of scientific apparatus including (by the 1680s) a microscope and a quadrant.[24] A project for a university museum in the early 1670s, on the other hand, fell through, as did plans for a philosophical society in the 1680s. Medical education was also quite well catered for, as at Oxford.[25]

At Oxford science even received official encouragement – following the establishment of the Savilian and Sedleian chairs early in the seventeenth century – when the Ashmolean Museum was built between 1679 and 1683 to house the collections that Elias Ashmole had brought together on the basis of the famous cabinet of the Tradescant family and had presented to the university.[26] The museum was a matter of some pride to the university authorities, and it attracted benefactions from virtuosi like Martin Lister and John Aubrey. Its first curator was Robert Plot, who used the laboratory in the building – 'perchance one of the most beautiful and useful in the world' – for chemical experiments and demonstrations. Plot had earlier been encouraged by the university authorities, who had armed him with a testimonial in his itinerant quest for natural historical information in the 1670s, and he was also one of the leading lights in the Oxford Philosophical Society, established in 1682.[27]

As the Ashmolean shows, support for science came from those

[23] Allen,[248] p. 245; B. Dickins, 'Henry Gostling's library: a young don's books in 1674', *Transactions of the Cambridge Bibliographical Society*, 3 (1961), 216–24.

[24] L. J. M. Coleby, 'John Francis Vigani, first professor of chemistry in the university of Cambridge', *Annals of Science*, 8 (1952), 46–60; Beale to Evelyn, 2 Jan. 1669, Evelyn, *Corr.*, no. 79; Shapiro,[250] p. 70n; D. J. Price, 'The early observatory instruments of Trinity College, Cambridge', *Annals of Science*, 8 (1952), 1–12.

[25] R. T. Gunther, *Early science in Cambridge* (Oxford, 1937), p. 20; Newton to Aston, 23 Feb. 1685, Newton,[127] II, 415; Rook,[249] pp. 117–20; Frank[132] pp. 207–11, 242–5; there are also relevant essays in Debus.[42]

[26] C. H. Josten, *Elias Ashmole* (Oxford, 1966), pp. 126–8, 205–6, 218–19, 232–4, 250–2; Mea Allen, *The Tradescants 1570–1662* (London, 1964).

[27] Wood, *Life and Times*, III, 54–6; Plot to Lister, 10 Feb. 1683, Gunther,[129] p. 366; Carr,[142] p. 29; Hunter,[143] p. 90; Gunther,[129] pp. 345–6, 354–7. See also above, pp. 55–6, 81–2.

at the heart of the Oxford establishment and was not associated only with outsiders. Even John Fell, Dean of Christ Church and Bishop of Oxford, who Aubrey thought 'influences all in all' there, considered adapting Tom Tower at Christ Church – then under construction – into an observatory in 1681, though dissuaded from this on the advice of Wren.[28] He also instigated the publication of ancient mathematical writers, and the press that he presided over brought out Plot's natural histories of Oxfordshire and Stafford-shire, Francis Willughby's *Historia Piscium* (1686) (on which Fell collaborated with the Royal Society), and the botanical work of Robert Morison. The latter involved the University in substantial expense, but enthusiasm for the project was displayed by a specimen put out in 1672, Morison's *Plantarum Umbelliferarum Distributio Nova*, which included engraved plates paid for by various Oxford dignitaries.[29] Other 'establishment' figures at Oxford with scientific interests included Obadiah Walker, who wrote a tract on optics, and Thomas Hyde, librarian of the Bodleian and an eminent Arabic scholar, who made microscopical observations of snails and insects and engaged in other natural historical studies.[30]

The links between Fell and Walker and opposition to the Royal Society have already been noted, and, paradoxically, almost all those involved in attacks on the Society at the universities and elsewhere could boast scientific interests. Peter Gunning was said to be a proficient mathematician who 'proved by Geometrie that there was a Deitie'; Robert South had some curiosity about science; Francis Turner had attended chemistry classes at Oxford in the 1650s; and Thomas Wharton had scientific commitments and connections, despite his tirade against the virtuosi.[31] Even Casaubon admitted that he was influenced by Bacon in his youth, performing experi-ments and inquiring about mechanical processes: he claimed that 'I did allwayes thinck the tyme as well spent in such communication, as if I had bine in my Studie, amongst all my bookes'. This is equally

---

[28] Aubrey to Wood, Dec. 1681, Bodleian Library MS Ballard 14, fol. 134; W. D. Caroe, *Sir Christopher Wren and Tom Tower, Oxford* (Oxford, 1923), pp. 31–2, 45–6.

[29] Stanley Morison, *John Fell, The University Press and the 'Fell' types* (Oxford, 1967), p. 39; Carter,[255] pp. 90–1, 236–9, 240. See also above, pp. 42, 69.

[30] [Walker], *Propositions Concerning Optic-Glasses* (Oxford, 1679); Hyde to Charle-ton, n.d., 3 Aug. 1680, Birch,[118] IV, 9–11, LBO VIII, fols. 121–7.

[31] Hunter,[143] p. 47; Jones,[245] p. 90n; Wood, *Life and times*, I, 472; Gillespie,[134] XIV, 286–7.

true of Stubbe: the sections of his books in which he tried to prove the incompetence of the Royal Society's investigations illustrate a wide scientific knowledge of which there is further evidence elsewhere, not least in his well-stocked library.[32] He also corresponded with Robert Boyle and wrote letters on natural historical topics to Sir Robert Moray, extracts from which were published in the *Philosophical Transactions*.[33]

How far, then, were attacks on the new science thinly veiled expressions of rivalry towards the Royal Society from supporters of other institutions with overlapping functions? Are they to be read merely as expressions of alarm at the Society's institutional existence and grandiose ambitions, surveyed in chapter 2? As far as the College of Physicians is concerned this seems certain, for the scientific subjects championed by the Society overlapped with the college's professional interests. In the Interregnum, after all, the college had been a centre of non-medical as well as medical research. Such evidence as survives suggests more of a demarcation of interest after the Restoration – perhaps particularly after the Fire of London, when the college was burnt down and its facilities destroyed – with only clearly medical activities there. The elegant theatre provided by the munificence of Sir John Cutler when a new building was erected after the Fire was intended for anatomical lectures, while the laboratory with furnaces and instruments to be found in the college in 1683 was used for experiments on mineral waters.[34] Reciprocally, the Royal Society sometimes tactfully held back on medical work, as when Wren suggested an anatomical committee in 1681 and Dr William Croone 'objected the college of physicians' (indeed, the quantity of activity at the Society's meetings connected specifically with medicine as distinct from the basic medical sciences was low throughout the Restoration).[35]

As Martin Lister discovered, there was a feeling in medical circles that doctors who wrote only on scientific topics had to prove their seriousness in medicine as well: apologising for publishing medical

---

[32] Rawlinson D.36.1, fols. 33–4; Casaubon,[265] pp. 21, 25; Stubbe, *Legends*, esp. pp. 35f, 89f; Sloane 35; Frank,[40] pp. 237–8; see also King,[41] pp. 167–72.

[33] Boyle,[71] I, xc–xcvi; Stubbe to Moray, 24, 30 Mar. 27 May 1667, EL S.1, fols. 89–91; *Phil. Trans.*, 2 (1667), 494–502, 3 (1668), 699–709, 717–22. Many of the views developed in Stubbe's published attacks on the Royal Society are, however, implicit in these letters.

[34] See above, p. 22 and also Webster,[73] pp. 411–12; Clark,[251] I, 330–1; Robinson to Lister, 15 Mar. 1683, Lister 35, fol. 89.   [35] Birch,[118] IV, 65; Frank,[106] p. 92.

research prematurely, he explained to Edward Lhwyd that 'I find my selfe necessitated to it, to stop the censorious mouthes, who thinke & say a man that writes on Insects, can be but a trifler in Phisic.' So tension existed, and the pamphlet dispute associated with Stubbe shows how it could flare up. This may also partly explain the disproportionate tendency of doctors who were once active in the Royal Society to leave (though professional commitments also often precluded scientific pursuits) and the failure of men like Sydenham and Wiseman to be proposed for membership despite the overlap of their interests with the Society's.[36] But it implied no hostility either in the college or the medical profession to the pursuit of natural know-ledge, as is clear from doctors' strong showing in non-medical science and the Royal Society throughout the period.

Likewise with the universities, Stubbe's and South's attacks on the Royal Society were connected with hostility to the Society's ambi-tions as a learned institution and with apprehensions about its likely interference in academic affairs. We have discussed Stubbe's warn-ing to the universities to watch out that the Society did not nominate candidates for preferment there and Aubrey's anxiety about Oxford's likely opposition to his educational scheme due to its challenge to the universities' academic monopoly. Similarly, Anthony Wood reported a rumour at Oxford that the Society 'desire to conferr degrees; the Universitie sticke against [this]', and Benjamin Woodroffe, Fellow of the Royal Society as well as an Oxford don, darkly remarked to a member of the Society's council in 1675 that 'I could wish none of that so worthily intended society had given occasion to us (who are the sworn members of another body) to be very wary how we concur in that there which may render us suspected here'.[37]

There was also a more general sense in which the Society's intellec-tual pretensions nettled some at Oxford and Cambridge. When defending the virtues of Oxford to a potential Royal Society critic, Ralph Bohun, Fellow of New College, protested: 'though I have an extreme respect to the modesty of their design, & faithfull delivery of experiments to the world, yet we can call them no more then a Learned Conventicle, if sett in Balance with the 2 most flourishing

---

[36] Lister to Lhwyd, n.d. [1693–4], Bodleian Library MS Ashmole 1816, fol. 176; Frank,[106] p. 97; see above, pp. 45, 47.
[37] See above, p. 126; Wood, *Life and times*, I, 354; Woodroffe to Southwell, 2 Mar. 1675, *H.M.C. Earl of Egmont*, II (1909), 36.

Universitys of Europe'. Comparable feelings undoubtedly underlay the hostility to the Society from Gunning at Cambridge and Fell at Oxford, and they were also encountered by the College of Physicians when it attempted to broaden its basis by electing honorary fellows in the early 1660s: for there was a rumour in Oxford and Cambridge that a move was afoot to infringe the privileges of the universities – though quite how is uncertain.[38]

Relations between the Royal Society and Oxford cannot have been helped by a wrangle over a valuable collection of books and manuscripts presented to the Society by the Duke of Norfolk, including various medieval, patristic and oriental manuscripts which both scholars at Oxford and prominent Fellows of the Society felt would be better off at Oxford, 'where they would remain more advantageously to all the interests of learning'.[39] It was planned that the Royal Society should be given useful mathematical and scientific books in exchange, thus suggesting an interesting division of labour along the lines proposed by Hooke when he defined the business of the Royal Society as: 'To improve the knowledge of naturall things, and all useful Arts, Manufactures, Mechanick practises, Engynes and Inventions by Experiments – (not meddling with Divinity, Metaphysics, Moralls, Politicks, Grammar, Rhetorick, or Logick).' Proposals to reallocate the Norfolk books were made in 1669, 1674 and 1677, but, in circumstances which remain rather shadowy, a strong party in the Royal Society resisted the bargain, evidently relishing the retention of volumes that Oxford wanted for the prestige of possessing them.[40]

From Oxford's point of view, the foundation of the Ashmolean may in part have been a similar gesture. This is certainly how it appeared to John Evelyn, who reported plans for the museum to John Beale as early as 1670, and the aggressive tone even in a relatively cautious scientific enthusiast helps to make clear Oxford's reaction: 'And yet when all is don, our *Alma Mater* men, you'l find, will be forc'd to play the Apes; You heare they talke already of founding a Laboratorie, & have beg'd the Reliques of old Tradescant, to furnish a Repositary, & will in time, civily invite the

[38] Bohun to Evelyn, 1668, Evelyn, *Corr.*, no. 301; Clark,[251] I, 312–14.
[39] Fell to Bathurst, 7 Mar. 1669, Evelyn,[126] p. 216.
[40] Sprat,[92] pp. 252–3; Birch,[118] II, 136, 351–2; Evelyn,[164] IV, 144–5; Weld,[94] I, 146; Evelyn to Fell, 12 Mar. 1669, Evelyn, *Lb.*, no. 314; Evelyn to Howard, 14 Mar. 1669, Evelyn,[126] pp. 217–18; Hooke to Bernard, 7 Apr. 1674, Smith 45, fol. 105. See also [William Perry], *Bibliotheca Norfolciana* (London, 1681).

despis'd Society, to vouchsafe a Correspondence, and for all their Antiquity, are but our Younger Brothers'.[41]

Those who thus resented the Royal Society felt no hostility to science. Bohun and Woodroffe, whose reservations have just been cited, could boast scientific interests like others involved in the dispute. Bohun had numerous scientific books in his library and wrote a scientific tract with Boyle's advice, which was commended by Oldenburg to his foreign correspondents; Woodroffe made careful provision for scientific teaching in the statutes that he drew up for the projected Worcester College at Oxford in the 1690s.[42] If anything, such men were likely to take umbrage at Hooke's suggested division of labour, which gave science to the Royal Society and other disciplines to the universities. Instead, those devoted to science at Oxford protested the harmony of such studies with those traditional in the university. Edward Bernard argued that 'books and experiments do well together, but separately they betray an imperfection, for the illiterate is anticipated unwittingly by the labours of the ancients, and the man of authors deceived by story instead of science'. John Wallis felt the same, trying to minimise the antagonism between the Royal Society and the universities. As he explained to Oldenburg:

Nor do I know that our designes do crosse one another. But I would not have you insist on that argument, (which you seem to lay some strokes upon,) that the University doth not meddle with Experimentall Phyolosophy. For it is a great mistake, (Experimentall Philosophy being as properly appertaining to the Constitution as any other; though, perhaps, in former times it have not been so much in fashion, as it now is here as well as with you:) You should rather say: It is no disparagement to the Universityes, for others to pursue philosophicall studies allso.[43]

So was the tension between the Royal Society and other centres of learning based only on institutional rivalry, whipped up by the extreme fears of eccentric reactionaries? Did it dwindle as the Royal Society proved itself a puny institution beside the universities or even the College of Physicians? On the contrary: such friction played its

---

[41] Evelyn to Beale, 27 Aug. 1670, Evelyn, *Lb.*, no. 329.
[42] Christ Church, Oxford, Evelyn MS 12; Bohun, *A Discourse Concerning the Origine and Properties of Wind* (Oxford 1671); Frank,[40] p. 239. Oldenburg to Cornelio, 9 Feb. 1672, Oldenburg,[121] VIII, 530; C. H. Daniel and W. R. Barker, *Worcester College* (London, 1900), p. 160.
[43] Bernard to Collins, 3 April 1671, Rigaud,[123] I, 158; Wallis to Oldenburg, 24 April 1669, Oldenburg,[121] V, 500.

part, but the debate ranged beyond the proper spheres of rival institutions to the broad claims that were made for the new science and their corollaries.

Significantly, almost all the published attacks on science date from a single, fairly short period around 1670 following the publication of the two defences of the new philosophy whose themes have already been summarised – Sprat's *History of the Royal Society* and Glanvill's *Plus Ultra*. These books made more explicit than any others the ideology of the new science in its widest implications, and, even if some of Sprat's statements were extreme and untrue, the stance that he and Glanvill took was broadly representative: the writings and correspondence of scientists are full of echoes of their optimism for the advancement of learning and their confidence in the utility and relevance of natural philosophy. Yet paradoxically these vindications of science by its eager propagandists did more than anything to cause the attacks that ensued. Casaubon's tract of 1669 was explicitly entitled *A Letter . . . to Peter du Moulin . . . Concerning Natural experimental Philosophie, and some books lately set out about it*, while Stubbe also found these books particularly alarming. Of Sprat he wrote: 'if that *History* take place, the whole education of this land and all religion is subverted', while it was felt at Cambridge that 'Mr. *Glanvill*'s books have much contributed' to the tension between the Royal Society and the universities.[44] What differences were involved?

The most widely-read exposition of Stubbe's and Casaubon's attack on Sprat and Glanvill, by R. F. Jones, places it in the context of the debate between 'Ancients and Moderns', the dispute as to whether authors in antiquity like Aristotle and Galen had reached an unequalled pinnacle in intellectual affairs or whether modern discoveries and inventions were bringing knowledge nearer perfection than ever before. This was undoubtedly an element in the episode. Championship of the moderns was writ large even in the title of Glanvill's book, and many scientific enthusiasts shared Bacon's progressivist view of intellectual achievement. As the Latitudinarian divine, Simon Patrick, put it: 'is it possible that so many new appearances should not alter the frame of Philosophy, nay rather hazard the pulling down of the old ruinous house that had too narrow foundations, that it may be built again with more magni-

[44] Stubbe to Boyle, 17 Dec. 1669, Boyle,[71] I, xc; du Moulin to Boyle, 28 Dec. 1669, ibid, VI, 579.

ficence?' Some went so far as totally to denigrate the works of the greatest scientific ancient, Aristotle, like the East Anglian virtuoso, Nathaniel Fairfax, who had 'rather be a Well-willer to a Brick-layer, than a Philosopher taking name from *Aristotles* Physicks'.[45]

In contrast, Casaubon regarded Aristotle as 'that inestimable magazin of human learning', believing 'that there is scarce any art, or faculty, wherein we do not come short of the Ancients'. John Fell was said to be 'wholly devoted to the old philosophy' and opposed to the study of moderns like Descartes, and the wide support for the scholastic synthesis at Oxford and Cambridge is clear from its vitality as the basis of teaching throughout this period.[46] With this faith in a fixed synthesis of established learning went mental attitudes that were also at odds with the new science and which the virtuosi therefore attacked – as when Glanvill had to argue at length against a hasty logical conclusion about optical instruments based on fallacious evidence. This was 'the *Notional* Way' that seemed so pervasive and unconstructive.[47]

The more attitudes towards the ancients are investigated, however, the less straightforward they seem. It is perhaps not surprising that Ralph Bohun had some very interesting arguments about the value of Aristotelian thought to understand the ideas even of those who had reacted against it. But even Glanvill and others associated with the Royal Society protested that it was not the whole of Aristotle that they objected to but only the '*notional* and *disputing* parts' and 'the swarms of Insectile Systemes and dilute Commentaries'.[48] John Wallis professed 'great Esteem' for the Stagirite philosopher, claiming that 'we have reason to think that if he had seen and observd some things which the latter times have discover'd, he would have, if not in some things changd his own Hypothesis at least added thereto'.[49]

Even if they abandoned the authority of Aristotle, many scientists took refuge in an appeal to other ancients, arguing anxiously that they were not innovating but rescuing knowledge lost at the Fall of

[45] Patrick,[91] p. 21; Fairfax, *A Treatise of the Bulk and Selvedge of the World* (London, 1674), sig. A5.
[46] Rawlinson D.36, fol. 26; Casaubon,[265] p. 29; Norman Sykes, *William Wake* (Cambridge, 1957), I, 10.   [47] Glanvill,[66] ch. 9.
[48] Bohun to Evelyn, 1668, Evelyn, *Corr.*, no. 301; Glanvill,[66] sig. B5 and chs. 1, 15, 17 passim; *Phil. Trans.*, 2 (1667), 412–13.
[49] Wallis to Middleton, 16 Sept. 1684, Society of Antiquaries of London MS 202, fol. 243.

Man or more recently. Isaac Newton drafted *scholia* for the *Principia* showing that the principles of gravitation had been known in antiquity: he echoed these more briefly in the published work, and it is clear that he regarded them not as literary embellishments but as an important part of his philosophy. Others tried to prove that Moses was the first atomist, while Hooke considered Ovid's views on the origins of the world as well as the evidence of fossils, seeing him as the heir to 'the most ancient and most knowing Philosophers among the *Ægyptians* and *Greeks*'. Indeed he went on to attack those who vilified the ancients, protesting the likelihood 'that in the more Youthful Ages of the World, there was a much greater Perfection of the Productions of it'.[50]

On the other hand, Stubbe was not really an 'ancient', which explains why R. F. Jones complained of his 'inconsistencies' when trying to force him into this role. At the start of his *Legends No Histories*, Stubbe apologised to those who might be surprised to find him on the side of antiquity – his 'modern' interests have already been mentioned – and his main concern seems to have been with the Royal Society's competence rather than the propriety of its activities: he was out to demonstrate how 'their *relations* and *Experiments*, are so *trivial, defective*, and false'.[51] As part of his assault on Glanvill's rhetoric he sought to readvance 'the Credit of the *Aristotelians*' and to prove that purely Galenic pathology was the basis of much useful practice, but he made no attempt to assert the unmatchable superiority of ancients over moderns. By illustrating how inflated were the Royal Society's claims and how much less it had achieved than Aristotle, he implied not that such attainment was impossible but that the new scientists were unoriginal, inept and dishonest: 'they either are *ignorant* of what is published by *Antient* and *Modern Writers*, or most *egregious plagiaries*'.[52]

So the question of 'Ancients and Moderns' was irrelevant to Stubbe, and, though Casaubon was an undoubted partisan, this issue was not central to his onslaught on the new science. Stubbe's and Casaubon's argument – which is briefly echoed elsewhere – was rather different. Theirs was essentially a debate about philistinism and its corollaries, an attack on the new science as a charter for the

[50] McGuire and Rattansi,[62] pp. 108f; D. B. Sailor, 'Moses and atomism', *J.H.I.*, 25 (1964), 3–16; Hooke,[65] pp. 379–80. See also Kubrin,[290] ch. 6, Teague,[27] pp. 122–3.
[51] Jones,[240] pp. 245, 253, 261–2; Stubbe, *Legends*, sigs. *1, *3.
[52] Stubbe, *P.U.*, title-page, sig. a2v, pp. 71f.; *Legends*, sig. *4v. Cf. King,[41] p. 168.

neglect of learned values which they saw as of the highest contemporary importance. As Stubbe put it, current fashion was turning against '*Antiquated Studies*' of classical and early Christian authors in favour of poetry, plays and '*Toyish Experiments*', which were claimed to be '*Cares* of such *high concernment*, that all *Philology* is but *Pedantry*; and *Polemical Divinity*, Controversies with which we are *Satiated*'.[53] This might seem borne out by Sprat's claims about the utility and pleasure of scientific studies for traditionally non-academic classes or by the anti-intellectualism of his opinion that the Royal Society's enterprise could 'be promoted by *vulgar hands*' with 'no extraordinary præparations of *Learning*: to have sound *Senses* and *Truth*, is with them a sufficient Qualification'. Here it is clear how scientists' aspirations to a broader role in society harmed them in the sphere where their reputation mattered most. Even Sprat's avowal (echoing Hooke's) that they only wanted to specialise in certain kinds of knowledge could backfire, for Sprat qualified it in a way that could be taken to imply hostility to other pursuits.[54]

It might be thought that Stubbe was only attacking Sprat and the fashionable hangers-on of science, but this is not true. Even the sober pages of Oldenburg's *Philosophical Transactions* trumpeted out attacks on those who 'preferre endlesse Contentions about Words before the usefull Works of the noblest Arts', observing that 'sometimes among Tradesmen and others *de plebe*, are found very intelligent and sagacious persons, excelling others, that have consum'd their whole life in public places of Learning'. Stubbe and Casaubon were right to see that iconoclasm was an essential part of the new science, a crucial factor in the intellectual revolution that Bacon and Descartes had wrought, and one strongly to be opposed for just this reason. There was a certain distrust of book learning even among major scientists: many of those committed to the new philosophy stressed that they had learnt more from men and real experiments than from books and that they were 'no great read men'. These included major scientists like Hooke and Wren as well as minor ones such as Francis Potter, and even if Hooke's vast library reveals a polemical element in such claims, it is not difficult to understand the anxiety they aroused.[55]

Such philistinism even antagonised groups fundamentally

[53] Stubbe, *Legends*, p. 11.   [54] Sprat,[92] pp. 435, 82–3. See above, p. 146.
[55] *Phil. Trans.*, 3 (1668), 630, 785; Hunter,[143] p. 40; Aubrey,[161] II, 164, 301; Evelyn to Wotton, 30 Mar. 1696, Evelyn,[126] p. 350. See above, p. 42.

favourable to the new philosophy, like the Cambridge Platonists. The divine, John Worthington, was alarmed at this tendency in Sprat, whom he called 'perfect Hylobares', obsessed by 'what gratifies externall sense, or what sense doth reach'. He thus echoed Henry More's earlier view that too great a concern for raising living-standards would produce 'a common wealth of rarely improved beastes, not of learned men'.[56] Casaubon went further, rightly tracing the iconoclastic basis of the new science to the attitudes of Descartes, whom More himself so much admired. Casaubon held that Cartesian influence had much to do with the decay of learning: he was unable to understand the fuss about the 'Cogito' principle – except for seeing its ominous implication for all existing knowledge – and inclined to write Descartes off as a charlatan. He was also unenthusiastic about mathematics and sceptical of its pretensions to revolutionise knowledge: he would have had no sympathy with those, like John Aubrey, who saw mathematics as the key to philosophy as a whole.[57]

Stubbe professed particular alarm at the threat of this philistine challenge in education. As he explained to the Earl of Arlington, Secretary of State, in defending his onslaught on the Royal Society:

I had observed that if the interest of our monarchy lay in preserving that education which was promoted in the universities, and that if the breeding of a succession of able statesmen, divines, and lawyers were a thing more than indifferent, my undertaking was generous and laudable. Ignorance is epidemical, and insensibly diffuses itself through the gentry and all professions, and common debauchery, atheism and popery have grown more than usually through our *virtuosi*.[58]

It is in fact difficult to find explicit attacks on the university curriculum among scientists beyond off-the-cuff remarks like Oldenburg's, just quoted. Most virtuosi seem to have agreed that the new philosophy, though important, was not the universities' main business and that traditional studies best prepared students for later life. Only extremists like John Beale advocated that science be made

---

[56] Worthington to More, 5 Feb. 1668, in R. C. Christie, ed., *The diary and correspondence of John Worthington*, II part 2 (Manchester, 1886), 265; More to Petty, 12 Mar. 1649, in Charles Webster, 'Henry More and Descartes: some new sources', *B.J.H.S.*, 4 (1969), 370.

[57] Rawlinson D.36.1, fols. 21–5, 32, 34; see also M. R. G. Spiller, 'The idol of the stove: the background to Swift's criticism of Descartes', *Review of English Studies*, new series 25 (1974), 15–24; Hunter,[143] p. 47.

[58] Stubbe to Arlington, 18 May 1670, *C.S.P. Domestic 1670* (1895), p. 225.

predominant so that 'then poor scholars would bring home to their parents some better skill, than to prove two eggs to be three'.[59] Though it is possible to find proposals for a new, scientific education in schools, both Sprat and Glanvill were careful to protest their loyalty to the universities and their disinclination to introduce major change there. They doubtless had memories of Interregnum reformers, and were only too anxious to avoid the smear of comparison with them. But – as so often – Sprat juxtaposed implicitly critical remarks with his otherwise unexceptionable comments on academic education, asking (in connection with schooling) 'Whether a *Mechanical Education* would not excel the *Methodical*?' and (with respect to teaching the gentry) if '*sweeter* and more *plausible Studies*' were not to be preferred to 'difficult, and *unintelligible Notions*'.[60]

These Stubbe used as a peg on which to hang a tirade against what he saw as the implication of Sprat's arguments for academic values, thus clearly revealing the priorities that he stood for. In his view, the universities

do enough in breeding up men to be fitly qualified for *Employments* in *Church* and *State*, and instruct them in so much *Philosophy* as is necessary for the explaining and defending of our Religion against *Atheists, Papists,* and *Socinians*: and whosoever shall put the *accurate debate* of *these Points, the Art of reasoning, the Validity of Consequences, the unfolding of critical Syllogisms and Fallacies, the general doctrine of Topicks*, the Moral Philosophy, and *Foundations of Civil Prudence*, (besides *Civil* and *Ecclesiastical History* and *Languages*) which are taught there, or ought to be by their *Constitution*, into the *Scales* on the one hand, and the *Mechanical Education* (recommended with all the advantages that ariseth from *Aphorisms of Cider, planting of Orchards, making of Optick Glasses, magnetick* and *hortulane Curiosities*) on the other hand, will be able to judge easily which *Studies* deserve the *most encouragement* by the *publick*, and which are most *useful* and requisite.

Stubbe accused the virtuosi of 'debauching our *Nobility, Gentry,* and all the *youth* from *those studies*, as *useless, empty,* and *impertinent*', and similar views were echoed by Casaubon and others.[61] In his own attack, Casaubon moved on from a critique of Sprat's

[59] E.g. Debus,[90] p. 244; *A Brief Vindication of the Royal Society*, p. 6; Wallis,[266] p. 315; Beale to Boyle, 24 Jan. 1667, Boyle,[71] VI, 422; cf., e.g., Henry Power, *Experimental Philosophy* (London, 1664), pp. 184–5.
[60] Glanvill,[66] sigs. B4v–5; Sprat,[92] pp. 59–60, 323f, 409. On school education see also above, p. 120.
[61] Stubbe, *P.U.*, p. 13; *Camp.*, p. 14. See also *Legends*, sig. *2. Casaubon,[265] pp. 24–5; Jones,[245] pp. 77–90.

and Glanvill's narrow, materialistic criteria of utility to an onslaught on their philosophical position and its pernicious religious corollaries. He had no doubt that science encouraged a stress on simplified and straightforward religion – the creed of the Latitudinarians which has already been referred to – and he regarded this with deep suspicion. His outlook reveals a genuine difference in intellectual priorities opened up by the new science and it is interesting that an attack on science was later juxtaposed with very similar aspirations by Dudley Loftus, one of the early members of the Dublin Philosophical Society.[62]

Casaubon, especially, illustrates a point of view which is often overlooked in the Restoration era. It is all too simple to stress the 'modern' features of the time which led to later developments, failing to question Latitudinarian clichés about '*rational Religion*' as the 'universal Disposition' of the age with science its natural support. Casaubon stands for a quite different viewpoint, completely ignored by one recent authority, espousing a learned, theologically and historically oriented view of the proper functions of academic life at odds with the priorities of science.[63] He was convinced that erudition was vital to the exposition and elucidation of Protestant Christianity and its defence against the Church of Rome and the threat of sectarianism, and there was arguably as much support for his opinion in Restoration England as for the attitude he attacked. The roots of this tradition are to be sought earlier in the seventeenth century, for the appeal to antiquity pervaded the whole of seventeenth-century Anglicanism. But the Interregnum sectarianism that encouraged the growth of natural religion among scientists also had a different effect, intensifying this stress on the authority of tradition as represented by Fathers and Councils of the early Church, and on the need for learned analysis of the scriptures and their historical background to guard against anachronistic misinterpretation.[64]

Casaubon was alarmed by such views as Sprat's assertion that 'now men are generally weary of the *Relicks* of *Antiquity*, and satiated with *Religious Disputes*', that religious controversy could be ended by an appeal to 'plain reason', and that religious knowledge required nothing 'but a bare promulgation, a common apprehension, and sense enough to understand the Grammatical meaning

[62] Casaubon,[265] pp. 4–6 and passim; Hoppen,[117] pp. 159f, esp. pp. 160, 164.
[63] See above, p. 30; cf. Cragg,[281] chs. 4–5.
[64] McAdoo,[87] chs. 9–10; Bennett,[254] esp. pp. 63–6.

of ordinary words'. No less shocking was Glanvill's conviction that the Bible was couched '*for the most part in the Language of Sense, being suted to Plebeian Capacities*', or his advocacy of comparative religion as a means of proving the validity of the Christian revelation.[65] These positions seemed to Casaubon either incredible or pernicious, and his estimate was echoed by Stubbe. Indeed, though scientists appealed to rationalism in religion as a defence against the excesses of the sects, Casaubon scented a link between the iconoclasm of their appeal to pure reason and the intellectual arrogance of the sectaries whom Sprat and others attacked: in an interesting passage in his unpublished tract he even accused Descartes of 'enthusiasm'.[66]

For men like Casaubon, tradition was the basis for moderation and compromise. Christianity had always had a controversial element and a heretical fringe, and there was therefore much of value in the teachings of the early Fathers and Councils on difficult topics, 'whose excellent works by a singular providence of God, are yet for the most part, extant'. These provided the basis of a definition of orthodoxy that had continued ever since, and it was essential for learned Protestants to be armed with a knowledge of them against the pretences of the Church of Rome. Stubbe was saying the same in more extreme form in characterising the new science as a Catholic plot to draw Protestants away from the defence of their church so that '*a Popish implicit faith*' could be introduced, and the plausibility of such views in the anti-Catholic atmosphere of Restoration England is easily underrated: others defended scholasticism as an equally essential controversial tool.[67]

In addition, Casaubon thought it essential to encourage 'generall learninge' in erudite languages, the history of rites and customs and the like, so that the text of scripture could be correctly understood and ignorant misinterpreters refuted. His ideal was a 'generall Schollar', a man

by whome doubts and difficulties of Scripture, must be resolved; the creditt and authoritie of the Scriptures themselves, as the Word of God, against all opposers mainteined; controversies in Religion, by the exact knowledge of former tymes, decided; forged wrytings and evidences, by which men, in all

---

[65] Sprat,[92] pp. 152, 22, 355; Glanvill,[66] pp. 141, 143–5.
[66] Casaubon,[265] pp. 17–18; Stubbe, *Censure*, pp. 40–52; Rawlinson D.36.1, fols. 22, 24 (cf. ibid, fol. 35 for Casaubon's allusion to Interregnum radical reformers).
[67] Rawlinson D.36.1, fol. 6; Casaubon,[265] p. 18; Stubbe, *Censure*, pp. 40, 53–5; *Camp.*, pp. 1–8; Barlow, *Genuine Remains*, pp. 155, 157–9. See Miller,[11] ch. 4.

ages, have bine abused and deluded, arraigned and convicted; the true meaning of doubtfull places in ancient authors, against the subtilties and impostures of Sophists, vindicated and asserted.

Casaubon described these skills in his manuscript tract, moralising both there and elsewhere on what he saw as their recent decay, which he attributed not least to the rise of experimental philosophy and the attendant view 'that all other learning is but verball and useless'. He was confident that such intellectual activity – 'true, solid, useful learning' – had an importance that science would always lack. Indeed, of science and technology he wrote: 'This kind of knowledge, if any man will cal it *learning*, I am not against it; and that it may be usefull for the better understanding of many places in Scripture, is not doubted.'[68]

This was no mere echo from the past. The intellectual activity that Casaubon extolled flourished in Restoration England, perhaps most spectacularly at Oxford, where it was associated with one of the most significant ventures of the period, John Fell's inauguration of an active scholarly printing house. Fell intended his press to champion the very programme that Casaubon outlined, though, of course, more general books were produced there as well. Its prospectuses (and, to a lesser extent, due to practical difficulties, its publications) show that Fell's first hope was to disseminate Biblical, patristic and early Christian texts – the materials for the learning that Casaubon advocated – and erudite works relating to them. Fell also patronised such research: he was the moving force behind John Mill's great *Novum Testamentum, cum lectionibus variantibus*, begun at this time though not published until 1707. In addition, he and other scholars like John Pearson, Master of Trinity College, Cambridge, and Bishop of Chester, produced major works of ecclesiastical scholarship which were admired throughout Europe, not least Fell's own edition of Cyprian.[69]

The chief representatives of this tradition were High Churchmen, and from the 1680s the appeal to antiquity increasingly became a party issue, its advocates becoming the 'High Flyers' and Non-Jurors of the early eighteenth century. But in the Restoration its support was wider: even the Latitudinarians commonly invoked the Fathers,

---

[68] Rawlinson D.36.1, fols. 2, 3, 34 and passim; Casaubon,[265] p. 2; Casaubon, *Of Credulity and Incredulity*, p. 296 and passim.

[69] Carter,[255] chs. 5–10; Hunter[256]; McAdoo,[87] pp. 386–7, 397–410; Fox,[257] p. 55 and pt. 2 passim; Bennett,[254] pp. 74–5; Sykes,[12] ch. 4.

and the appeal to antiquity – judiciously balanced with reason – was quintessentially Anglican.[70] The importance widely attached to Biblical scholarship is shown by the alarm that greeted the appearance in English in 1682 of the *Critical History of the Old Testament* by the unorthodox French divine Richard Simon, which threw doubt on accepted views about the transmission of the text of scripture. Even a layman like John Evelyn was anxious to see a learned reply to this, who, 'though he be no man of the Church, is yet a son of the Church, and greatly concerned for her; and though he be not learned, he converses much with books, and men that are as well at Court as in town and the country'. He was sure that a learned reply was needed every time a heterodox work came out, since 'the men of this curious and nicer age do not consider what has been said or written formerly, but expect something fresh'.[71]

One can extrapolate further from Casaubon's attack on science from the viewpoint of contemporary learned values. Though he was concerned mainly with the religious functions of erudition, his devotion to accurate scholarship had its counterpart in a related antiquarian tradition which was equally lively and served similar ends: this too, it may be noted, was encouraged by Fell.[72] In the work of men like Sir William Dugdale, Robert Brady and Henry Wharton, Restoration England could boast a remarkable achievement in editing historical texts, investigating the origins of rights and privileges, studying ecclesiastical antiquities and elucidating the constitutional history of medieval England. As is clearest from the heated historical controversy on constitutional issues (though it was also true of ecclesiastical antiquarianism), this too was closely related to matters of high contemporary priority. It was the learned extension of political debate and its protagonists were often lawyers, politicians and churchmen whose scholarship informed their and their colleagues' practice.[73]

This vigorous research has sometimes been compared with the new science, but it was essentially separate from it and potentially antipathetic (though no antiquary attacked science as those devoted to ecclesiastical learning did). Their different priorities are shown by

[70] G. Every, *The High Church party 1688–1718* (London, 1956), esp. ch. 1; Patrick,[91] pp. 9–10; Bennett,[254] pp. 72–3 and passim; McAdoo,[87] pp. 387–8, 399.
[71] L. I. Bredvold, *The intellectual milieu of John Dryden* (Ann Arbor, 1934), pp. 102–6; Evelyn to Fell, 19 Mar. 1682, Evelyn,[126] pp. 265–6. Cf. e.g. Hoppen,[117] p. 161; Heath to Evelyn, 1667, Evelyn, *Corr.*, no. 881 (cf. Evelyn,[164] III, 502 n. 5).
[72] Carter,[255] pp. 227–8; Hunter,[256] p. 526.   [73] Douglas[258]; Pocock,[259] chs. 8–9.

the contrast between the antiquarian interests of scientists and of those at the centre of the historical tradition (beside whom the 'scientific' antiquaries were in a distinct minority). Virtuosi like John Aubrey, Robert Plot and Martin Lister neglected the learned elucidation of texts in favour of archaeological antiquities in the field, studying '*things*' as against '*persons* and *actions*'.[74] Their preoccupation with understanding the natural world and technological process rather than controversy and precedent reveals a split in aim and method similar to that manifested by the attacks of Casaubon and others on the priorities of science. They also show the justice of the accusation of philistinism levelled at the new philosophy. In comparison with their learned contemporaries, these archaeological antiquaries were notable for their sheer ignorance, well illustrated by Aubrey's querulousness about the basis of some of his novel archaeological theories. It was partly because they were unfamiliar with erudite sources that they reached conclusions from material relics of the past in isolation from them, whereas those who tried to collate the two were frequently misled. Paradoxically, modern archaeologists respect these pioneers for the very characteristics that learned contemporaries despised in them.[75]

It would be artificial to seek a complete polarisation of science and learning. There was no absolute gulf between history and archaeology, as there were overlaps between the protagonists of science and of ecclesiastical and historical scholarship. John Wallis was not only an active scientist but also 'no mean Critik in the Greek and Latin Tongues', and others who combined active learned and scientific interests included Edward Bernard and Thomas Gale. Bernard played an important part in encouraging Biblical scholarship as well as science at Oxford, and his name is associated with a major catalogue of antiquarian manuscripts that was compiled under his supervision and published in 1697. Gale, successively Professor of Greek at Cambridge, High Master of St Paul's School and Dean of York, was not only an erudite classical and historical scholar but also an active officer of the Royal Society in the 1670s and 1680s. Patrons like the statesman Sir Joseph Williamson also found a place for both learned and scientific projects.[76]

[74] Plot,[68] p. 392; Hunter,[143] pp. 25–6, ch. 3.
[75] Hunter,[143] pp. 224–5; Levine,[261] chs. 8–9.
[76] T. Hearne, *Remarks and collections*, ed. C. E. Doble, 1 (Oxford, 1885), 198; T. Smith, *Vita Edwardi Bernardi* (London, 1704); Fox,[257] pp. 60–1; Douglas,[258]

But this does not detract from Casaubon's percipience in seeing the latent conflict between the two schools, and subsequent developments were to justify his apprehensions. Within half a century there was a noticeable decline in the polemical research of historical and ecclesiastical scholars, which was mirrored by a dwindling in the authority of the Fathers in theological circles.[77] These were offset by the popularity of a less erudite (often archaeological) antiquarianism, and by the triumph of religious opinions of the kind associated with the Latitudinarians in the context of a burgeoning of science. It was part of a more general shift in European values, that 'Crisis of the European Conscience' that has long been familiar, in which historical argument decayed and rationalistic notions came increasingly to the fore.[78] In the Restoration this development was in embryo, but it underlay even apparent unanimity. Thus though the Latitudinarians respected the Fathers, they tended to stress their rationality rather than their antiquity. And though Boyle argued in terms not so different from Casaubon's about the need for linguistic and historical erudition to properly understand the Bible, it is symptomatic that he reversed Casaubon's priorities: it was learning that he had to justify to his audience while taking the study of nature for granted as a Christian duty.[79]

So the new philosophy apparently had greater staying power than the learned traditions with which it co-existed. In the late seventeenth century, however, compatibility was inevitable, not least since science and learning were natural allies in another controversy about the proper functions of intellectual life which came to a head in the 1690s. This was the notorious debate of 'Ancients and Moderns' in which the chief participants were Sir William Temple and William Wotton and which was very different from the quarrel between Stubbe and Casaubon and the virtuosi: indeed it is difficult to see why R. F. Jones conflated the two in his famous account of seventeenth-century intellectual history.[80]

At the heart of this later debate was a much older argument

---

pp. 59–61, 66, 111–12 and passim; Birch,[118] III 450f, IV, passim; Allen,[231] pp. 64f; Hunter,[107] p. 21, F101; Carter,[255] pp. 50–2, 53.

[77] Douglas,[258] ch. 13; Bennett,[254] pp. 73–4, 76; Duffy,[285] pp. 298–300.

[78] Hunter,[143] pp. 202–5; S. Piggott, *Ruins in a landscape* (Edinburgh, 1976), chs. 6–7; Hazard[334]; and see below, pp. 191–3.

[79] Bennett,[254] p. 72; Patrick,[91] pp. 10–11; Boyle,[71] II, 257–60, IV, 63; ibid, IV, 37.

[80] Jones,[240] pp. 266–7.

between rhetoric and philosophy which had its roots in classical antiquity and had received a new boost during the Renaissance.[81] Was it more important to cultivate virtue and eloquence, or to pursue accurate knowledge of the natural world and the human past? Since the ancients had excelled in certain subjects of 'moral' utility, their superiority was clear whatever was said of modern achievements in science or erudition. In other words, here was a frontal assault on the priorities of both sides in the earlier dispute, since both were dependent on the cumulative progress of knowledge, even if they differed in their view of which kind of learning was more important.

It is possible to find hints of this quarrel in England before the 1690s, not least concerning the proper content of education. Thus Obadiah Walker, writing *Of Education, Especially of Young Gentlemen* (1673), had no doubt that the first priority of schooling was to cultivate wisdom as against knowledge: 'the most useful knowledge is that, of a mans self', he argued, considering it best for a man to excell in 'virtue, prudence, and those abilities which render him more useful in the generall concernments of Mankind'. Hence he saw too great an emphasis on science as a hindrance, and others in the Interregnum and Restoration argued similarly for virtue rather than practicality as the proper end of education.[82] To some extent, fashionable attacks on science like those of Charles II, Butler and Shadwell reflect similar humanist ideals, and Sprat, in his *History of the Royal Society*, tried to reply to those who disparaged science 'as that which disables us, from taking right measures in humane affairs'.[83]

Temple, however, made the theme more explicit than ever before in his 'Essay upon the Ancient and Modern Learning' (1690), as he expounded his esteem of genius as against accumulation, of eloquence against scholarship, of knowledge of men as morally superior to knowledge of books. Temple was as hostile to Casaubon's ecclesiastical erudition – 'the humour of ravelling into all these mystical or entangled matters' – as he was to the pretensions of the new science. He saw any such painstaking research as inferior to the wisdom attained by the ancients and emulated by their latter-day devotees and hence argued that all the achievements of the moderns

[81] Levine,[261] pp. 19–21; and see also his forthcoming study of the whole affair.
[82] Walker,[268] preface, p. 112; Jones,[245] p. 79.
[83] See above, pp. 70, 131; Sprat,[92] p. 27.

were overrated.[84] In the aftermath of Temple's attack his supporters lampooned the inconclusive triviality of science in the satirical *Transactioneer* just as they pilloried research in the famous Phalaris affair, in which exact modern scholarship was pitted against the erroneous self-confidence of the ancients.[85]

The opposing case was put by William Wotton, who brought together the achievements of the moderns in both science and erudition in his *Reflections upon Ancient and Modern Learning* (1694). Though generally optimistic about the progress of knowledge and opposed to the adulation of anything 'because it is *oldest*, not because it is *best*', Wotton attempted a judicious compromise, separating out topics where knowledge was cumulative from those in which it was not. Thus the ancients excelled in their 'acuteness of wit and elegancy of language', as John Ray put it in a succinct summary of Wotton's argument, and their moral and artistic prowess was at best only equalled by the moderns. But in numerous disciplines the moderns had undoubtedly surpassed their precursors, including philology, chronology, geography and all the sciences.[86]

This compromise increasingly prevailed, and after this climax in the 1690s the quarrel dwindled, except in so far as it reflected perennial issues about the priorities of intellectual life which have outlived the episode as they preceded it. Science did not dominate the controversy: it was important largely because its Baconian stress on the cumulative progress of knowledge brought the old dispute to a new climax. In terms of the intellectual status and implications of natural philosophy this was arguably a subordinate episode. It was certainly less significant than the passionate debate earlier in the Restoration about science, learning and iconoclasm.

[84] Temple,[267] p. 66 and passim.
[85] Levine,[261] pp. 247–9; Levine,[263] pp. 53–4; Jones,[262] p. 36.
[86] Wotton,[269] sig. A6v and passim; Ray to Robinson, 15 Dec. 1690, Ray,[124] p. 229; Levine,[263] pp. 47–9.

7

Atheism and orthodoxy

mmmmmmmmmmmmmmmmmmmmmmmmmmmmmmmmmmmmmmmmmmmmmmmmmmmmmmmmmmmmmmmmmmm

In considering the social influence of seventeenth-century science, historians have often been guilty of a marked anachronism. They have frequently assumed that the most important way in which science could affect society was through the application of inventions and improvements to the amelioration of human life, and have studied science's setting accordingly. But, as we saw in the last chapter, numerous contemporaries felt that the priorities of intellectual activity should be quite different and thought the new science pernicious precisely for this reason. For many, the most important social corollary of intellectual life was the way it informed conduct. It was widely believed that the fabric of society depended on a philosophical and theological consensus, and hence unorthodox viewpoints which seemed to undermine this and to encourage immoral attitudes were regarded with alarm, their refutation or suppression being seen as a matter of great moment.

Seventeenth-century Englishmen tended to express these anxieties by speaking of 'atheism', and paranoia about this threat was as pervasive as the hate of Roman Catholicism which was so powerful in politics and which is only intelligible after careful scholarly analysis. Anti-popery has now been sympathetically studied, but the fear of atheism – which shows an equally revealing sense of insecurity – still awaits satisfactory treatment.[1] It is therefore essential to describe the phenomenon before illustrating its equivocal implications for the new science.

Contemporaries undoubtedly took this threat very seriously. Ralph Cudworth, the Cambridge Platonist, spoke for many when answering those who questioned whether infidelity was as great a

---

[1] Miller,[11] esp. ch. 4; K. H. D. Haley, ' "No Popery" in the reign of Charles II' in J. S. Bromley and E. H. Kossmann, eds., *Britain and the Netherlands*, v (The Hague, 1975), 102–19. For general accounts of atheism see below, p. 214.

menace as he claimed: 'we could heartily wish upon that condition, that all this labour of ours were superfluous and useless. But as to Atheists, these so confident exploders of them are both unskilled in the monuments of antiquity, and unacquainted with the present age they live in; others having found too great an assurance, from their own personal converse, of the reality of them.' Nor was this simply a clerical rearguard action in an increasingly secular age. In the Restoration, 'discourse about Divinity [was] the frequentest table-talk in *England*' and religious books sold far better than any others, even if a growing defensiveness can be discerned in them.[2] Numerous laymen were concerned about heterodoxy, as will be seen, and though the differing anxieties of heresy-hunters and the varying inclusiveness of their accusations would repay precise analysis, this has yet to be undertaken.

Strictly interpreted, atheism meant, in the seventeenth century as in the twentieth, denial of the existence of God. There was debate at the time – as there has been more recently – as to how common atheists in this sense were.[3] But it does not altogether matter, since atheism was usually taken to include not only complete disbelief but also any doubt of an active deity constantly at work in the universe, on the grounds that the two were inextricably related. Francis Gastrell, one of the numerous ecclesiastics who assailed atheists, included in the category 'whoever holds such an Opinion which *exempts him from all Obligation of Duty to a Superiour Being*, or *cuts off the Expectation of Rewards and Punishments* consequent thereupon': it was widely believed that without this the moral status quo was as much at risk as from absolute incredulity. Gastrell was even suspicious of anyone whose religion comprised less than 'that whole Scheme of Humane Duties we find delivered in the Writings of the New Testament, as recommended and inforced by such a future State as is there described'.[4]

This is a salutary reminder that belief in an active – potentially arbitrary – providential God retained immense importance for religious orthodoxy in late seventeenth-century England. It is easy to

2 Cudworth,[314] I, xlvii; Walker,[268] p. 218; Sommerville,[282] pp. 29–30, 115 and ch. 7 passim; see also C. J. Sommerville, 'Religious typologies and popular religion in Restoration England', *Church History*, 45 (1976), 32–41.

3 Lucien Febvre, *Le problème de l'incroyance au sixième siècle* (Paris, 1947), esp. conclusion; Allen,[270] pp. 2–3 and n.4.

4 Bentley,[316] III, 3–4; Gastrell, *The Certainty and Necessity of Religion in General* (London, 1697), pp. vii, 251.

assume that the movement towards Deistic religion was more advanced than it actually was and to neglect this crucial, voluntarist streak which underlies much of the anxiety about irreligion as thus broadly defined.[5] The infidelity of contemporary England seemed to many contemporaries to presage not just possible social breakdown but the inexorable wrath of God, who might suddenly visit the land with 'some signal judgement' (as John Evelyn reflected in 1699), if not the fate of Sodom and Gomorrah.[6]

Not only did people conflate into atheism denial of an active providence. Atheism was also commonly taken to include attitudes and behaviour that were thought to accompany an irreligious outlook: there was as much concern about 'practical' as 'philosophical' disbelief. Almost the only man publicly humiliated for his godless views, Daniel Scargill, a Cambridge don, was forced to admit that 'agreeably unto which principles and positions, I have lived in great licentiousness, swearing rashly, drinking intemperately, boasting my self insolently, corrupting others by my pernicious principles and example'. The scholar and divine, Richard Bentley, was convinced that sermons against infidelity had to be '*contra malos mores*, not *malos libros*', against bad habits, not bad books: 'Atheism is so much the worse that it is not buried in books; but is gotten εἰς τὸν βίον [into life], that taverns and coffee-houses, nay Westminster-hall and the very churches, are full of it.'[7]

The Restoration saw as much devotion to the reformation of manners as to the refutation of heterodox ideas, and in the 1690s a number of societies were set up to deal with the practical side of the problem that 'atheism and profaneness' were causing. Many argued that practical infidelity was the root of the trouble. 'Atheism begins', the future Archbishop of Canterbury, Thomas Tenison, opined in 1691, 'not from the Arguments of a sound Mind in a sober Temper, but in a sensual Disposition, which inclines Men to seek out for colours, whereby they may deceive themselves into an opinion of the safety of living in a course that pleases them.' It was nevertheless

---

[5] Thomas,[57] ch. 4; Skinner,[109] pp. 130–1.

[6] Evelyn,[164] v, 366; Robert South, *Sermons Preached upon Several Occasions*, new edition (London, 1823), I, 374.

[7] *The Recantation of Daniel Scargill* (Cambridge, 1669), pp. 3–4; see also J. L. Axtell, 'The mechanics of opposition: Restoration Cambridge v. Daniel Scargill', *Bulletin of the Institute of Historical Research*, 38 (1965), 102–11; Bentley to Bernard, 28 May 1692, in *The correspondence of Richard Bentley*, ed. C. Wordsworth (London, 1842), I, 39.

believed that philosophical positions buttressed this and hence deserved attack: even if 'Immorality is the beginning of Atheism', 'Atheism is the strengthning of Immorality'.[8]

At the most extreme, atheists were thought to threaten a complete breakdown of public order due to their irresponsible and selfish attitudes – a point of view that was sometimes rhetorically illustrated by conflating them with republicans. Tenison felt that godless principles necessarily denied the validity of oaths and hence supplanted all laws: 'It behoveth all *Civil Governours* to animadvert upon Atheism, as that which supplants the Foundation of all Humane Society.' If the impious came to outweigh the pious, 'we should soon become a *Desolation*', while Joseph Glanvill saw the '*quibbling debauches*' with whom he associated ungodly attitudes as the enemies of government and religion of any kind.[9]

Hence 'atheism' is a very diffuse phenomenon, the more so because it could easily be extended to include views and attitudes that were thought – if only by implication – to lead in an irreligious direction. For by throwing about the smear of atheism people identified what they found disquieting in their age, even if it seems ludicrous in retrospect thus to label the attitudes and activities involved. This partly explains the uncertainty about the prevalence of infidelity. Whatever the situation about cabals of 'downright and professed Atheists' – for whom Cudworth, though convinced of their existence, had little hope – there were also thought to be many moderates. In any case, one aim of the anti-atheist literature was to confirm the faith of 'weak, staggering, and sceptical Theists' and to enable all believers to 'effectually learn how to purge our Minds of those ill Qualities which naturally are subservient to *Atheism*'. The Puritan divine, Stephen Charnocke, even saw atheism as a natural and universal occurrence among Christians, comprising any departure from proper Godly commitment.[10]

---

[8] Bahlman,[284] p. 1 and passim; see also Duffy,[285] pp. 293–5; and T. C. Curtis and W. A. Speck, 'The societies for the reformation of manners: a case study in the theory and practice of moral reform', *Literature and history*, 3 (1976), 45–64. Tenison, *A Sermon Concerning the Folly of Atheism* (London, 1691), pp. 15–16.

[9] E.g. Prideaux to Ellis, 11 Dec. 1693, in *Letters of Humphrey Prideaux to John Ellis*, ed. E. M. Thompson (London, 1875), pp. 162–3; Tenison, *A Sermon*, pp. 32–5; Glanvill, *A Blow at Modern Sadducism* (London, 1668), p. 147.

[10] Cudworth,[314] i, xlvii–xlviii; John Edwards, *Some Thoughts Concerning the Several Causes and Occasions of Atheism* (London, 1695), preface, p. 128; Charnocke, 'A Discourse upon Practical Atheism' in *Several Discourses upon the Existence and Attributes of God* (London, 1682), pp. 47–108.

In religious terms, atheist accusations could thus express hostility to the appeal to basic beliefs and natural theology considered in the last chapter. Many thinkers who tried to do good by cutting religion down to its essentials found themselves attacked as infidels for their pains, including not only John Locke and those writers called 'Deists', whose actual views were often caricatured by their opponents, but also as conformist a figure as John Wilkins.[11] Even antagonists of unbelievers who sought 'to argue from common Principles, which they cannot disavow', on the grounds that 'the weight of Authority is of no force with *Libertines*', risked being accused of a dangerous sell-out to heterodoxy.[12]

Similarly in morality, complaints about 'practical atheism' often affirmed concern about decaying standards. Accusations of irreligion expressed anxiety about 'those destructive vulgar vices, the contagion of which is become epidemicall' even among the gentry in the provinces, quite apart from their inroads among the impressionable undergraduates of Oxford.[13] Above all, disquiet was felt about the libertinism of high society and the court, which seemed all the more dangerous because lax moral standards at so elevated a level were bound to percolate throughout society. The culture of the Earl of Rochester and the court rakes, of William Wycherley and the grosser of Restoration comedies, though often seen as characteristic of the age, aroused grave reservations among many. John Evelyn, whose courtly connections brought him into contact with it, felt nothing but scorn for 'these *magnificent Fops*, whose *Talents* reach but to the adjusting of their *Peruques*, courting a *Miss*, or at the farthest writing a smutty, or scurrilous *Libel* (which they would have pass for *genuine* Wit)'.[14]

Of all features of the age which seemed alarming and naturally concomitant to atheism, none was more often attacked than the satirical flavour of fashionable literature and conversation. Sir William Temple, in his 'Essay upon the Ancient and Modern Learning', deprecated 'the vein of ridiculing all that is serious and good, all honour and virtue, as well as learning and piety', which he considered 'the itch of our age and climate'. In a sermon delivered before the King in 1675, John Fell went further, seeing this propensity to

---

[11] Yolton,[293] ch. 1, pp. 169f and passim; Colie[277] esp. p. 38; Skinner,[109] p. 226.
[12] William Bates, *Considerations of the Existence of God* (London, 1676), sig. A6.
[13] Cole to Plot, 25 Mar. 1685, Gunther,[129] p. 285; Anthony Wood, *Life and times*, II, 95–6, 125.   [14] Evelyn,[197] sig. A1v.

scoff at all worthwhile attitudes and activities as more widespread than ever before and hence heralding the end of the world, as foretold in the Second Epistle of St Peter:

That this Age of ours has somwhat of mockery for its particular Genius, so that scarce any thing is so entertaining, as to sport with the misadventures or failances of others; nor no faculty more recommending then the being dextrous in turning serious things to Ridicule, I think is a truth so notorious, that I may say it without offence to any.[15]

Such flippant and sarcastic attitudes were frequently linked to infidelity. John Wallis spoke of those who 'thought it *a piece of Wit* to pretend to *Atheism*', and it was noted that atheists tended to be 'a sort of Jesting, Scoffing People, giving themselves to Railery and Burlesque'.[16] Not only the sceptical and dismissive tone of satire aroused reservations, but also the confident rationalism to which it gave expression. There is a link between philosophical unorthodoxy and the increasingly rational tone of political debate in the later seventeenth century, while the growing scepticism of the intelligentsia towards witchcraft, most recently chronicled by Keith Thomas, was attributed at the time to nascent irreligion among 'the looser *Gentry*, and the small pretenders to *Philosophy* and *Wit*'. Most telling, perhaps, of the disquiet about the age's secularising tendencies that atheist accusations expressed was John Fell's challenge to those who lived dissolutely and defied God to 'lay their hands upon their breast, and ask themselves at what ensurance office they have secur'd a longer date of life'.[17]

The roots of all these interrelated attitudes were commonly traced to the Interregnum. The breakdown of order in the Puritan Revolution and its aftermath was blamed for the decay of moral values – John Aubrey saw the drunkenness and rudeness of his country neighbours as 'dregs of the Civill warre' – while the proliferation of sects was frequently blamed for weakening the religious consensus and encouraging scepticism. To some extent it was thought that sectarianism led to atheism, and stories were told of men who ran the gamut of extremist religious groups before ending as unbelievers.

[15] Temple,[267] p. 70; Fell, *The Character of the last Daies* (Oxford, 1675), pp. 15–16, 19.
[16] Wallis, *Hobbius Heauton-timorumenos* (Oxford, 1662), p. 6; Edwards, *Some Thoughts*, p. 29.
[17] Skinner,[275] pp. 294–7; Thomas,[57] ch. 18; Glanvill,[315] p. 62; Fell, *The Character of the last Daies*, p. 23.

But equally, religious indifference was thought to be a natural reaction to 'enthusiastic' excesses: it was not least to 'the *spiritual vices* of this *Age*' that Thomas Sprat attributed the fact 'That the influence which *Christianity* once obtain'd on mens minds, is prodigiously decay'd'.[18]

The Interregnum had also seen a ferment of ideas, and the new currents of free thought strengthened the philosophical basis of atheism, particularly through the increased currency of a materialism which denied divine intervention in the world. To some extent ancient theories were revived: Lucretius received attention even from as orthodox a figure as John Evelyn, who disseminated materialist ideas despite his reservations about them. But these were reinforced by the new doctrines of Descartes and Hobbes, and, while Descartes's position was ambiguous, causing problems for scientists that will be examined in a moment, fewer reservations were felt about Hobbes. For Hobbes also expressed other positions thought to abet atheism, including a critical attitude towards the Bible, while his *Leviathan* was thought to exemplify the political and social implications of infidelity.[19]

Indeed, Hobbes became the bogeyman of the age. His works were burnt by the public hangman, proscribed by the University of Oxford, and attacked by numerous devout polemicists who held him responsible for the 'Hobbist' views which they saw about them. In the course of this Hobbes suffered a good deal of travesty, for he claimed to be (and can be shown to have been) perfectly religious, if far from orthodox.[20] Much of what his antagonists disliked and loosely described as 'Hobbism' was a more generalised scepticism which did not necessarily owe much to Hobbes, whose popularity was arguably due to the commonness of a mentality to which his ideas appealed. John Aubrey's 'Life of Mr Thomas Hobbes of Malmesbury' was written partly to prove that his friend was not a 'Hobbist', while Hobbes suffered from the prevalent 'scoffing humour' himself.[21] But nevertheless Hobbes's was the name most frequently associated with atheism – Hobbes was in genuine fear of

---

[18] Hunter,[143] p. 217; Glanvill, *A Blow*, p. 158; Sprat,[92] p. 376.

[19] Evelyn, *An Essay on the First Book of T. Lucretius Carus, De Rerum Natura* (London, 1656); Christ Church, Oxford, Evelyn Collection MSS 33–4; Mayo,[273] ch. 3; Hobbes,[70] III, pt. 3 and passim.

[20] Hunter,[143] p. 55; Mintz,[274] pp. 61–2 and passim; W. B. Glover, 'God and Thomas Hobbes' in Brown,[276] pp. 141–68.

[21] Mintz,[274] pp. 136–7; Skinner,[275] pp. 294–7; Aubrey,[161] I, 340 and 321–77 passim.

prosecution for heresy after the Restoration – and he was considered the chief source of irreligious notions. As Richard Bentley put it in 1692: 'There may be some Spinosists, or immaterial Fatalists, beyond seas. But not one English Infidel in a hundred is any other than a Hobbist; which I know to be rank Atheism in the private study and select conversation of those men, whatever it may appear to be abroad.'[22]

As Bentley's words suggest, this emphasis on Hobbes oversimplified the philosophical position. Mechanistic materialism like Hobbes's was seen as the main strand of heterodoxy but danger was also felt from pantheism, from the view that God and nature were the same, that God was himself diffused through matter. This opinion was especially associated with Spinoza, though a few English (and ancient) precedents can be traced, and it naturally caused alarm because of its implication that God was divisible and expendable.[23] More or less extreme ideas of this kind were espoused by the early English Deists, including the iconoclastic and incendiary Charles Blount, while a variation on the theme is found in the cosmology of John Toland, author of *Christianity Not Mysterious* (1696): Toland had absorbed a mixture of heretical views, including those of the sixteenth-century thinker, Giordano Bruno.[24]

It would be mistaken, however, to separate 'Hobbesian' and 'Spinozist' materialism too carefully, for they were frequently conflated by the guardians of orthodoxy, who were inclined simple-mindedly to write off ideas like Spinoza's as extreme versions of Hobbes's. Similarly, though Toland's debt to Bruno has been asserted by a modern scholar, to Leibniz he seemed like a Hobbist or a Lucretian.[25] It is dangerous to subject the doctrines that contemporaries found alarming and labelled as irreligious to too close philosophical scrutiny, for this disguises the diffuse nature of the atheist menace and the fact that it was only partially metaphysical at the best of times. Bluntly mechanistic materialism like Hobbes's might on the face of it seem mutually contradictory of pantheism,

[22] S. I. Mintz, 'Hobbes on the law of heresy: a new manuscript', *J.H.I.*, 29 (1968), 409–14; Bentley to Bernard, 28 May 1692, in *The correspondence of Richard Bentley*, 1, 39–40.

[23] Cudworth,[314] 1, xl–xli, 215–16; Colie,[278] esp. p. 186; Jacob,[301] p. 230 n.7. See also his forthcoming study of Stubbe and Jacob.[308]

[24] Colie,[277] p. 37 and passim; Redwood,[279] pp. 494–5; Jacob,[280] pp. 309, 316f.

[25] Colie,[278] esp. pp. 183–5, 203–4; Mintz,[274] pp. 57–9; Colie,[296] ch. 5; Jacob,[280] pp. 315–16.

and Cudworth believed that if left to themselves they would destroy one another: yet this did not reduce his conviction of the reality of the danger and the need to combat it.[26] Indeed the vagueness of usage made the problem greater rather than less, exacerbating difficulties stemming from the usage of atheism to identify disquieting features of contemporary thought that has already been noted: for this loose definition was juxtaposed with real fears about a serious social threat.

All this should illustrate how vulnerable science was to atheistic accusations. The new philosophy was closely associated with the stress on simplified, rational religion, while its rise to popularity in the Interregnum had coincided with the flourishing of heterodox ideas that was so widely deprecated. In addition, the iconoclastic implications of the new science disturbed many, and, as shown in the last chapter, these too were often described as 'atheistic' in a loose, disapproving sense. The chief difficulty, however, was scientists' preoccupation with secondary causes, encouraged by Bacon's influential distinction between the proper spheres of science and religion. Ralph Cudworth found in Bacon's rejection of final causes 'the very spirit of atheism and infidelity', and the divine John Edwards echoed him in the 1690s: 'Learned Enquirers are apt to give Encouragement to Atheism by *an obstinate endeavouring to solve all the* Phoenomena *in the world by mere Natural and Corporeal Causes*, and by their averseness to admit of the aid and concurrence of a Supernatural or Immaterial Principle for the production of them.'[27]

To some extent science was at risk even in its milieu, in so far as it shared the fashionable London circles where libertine and 'scoffing' values flourished among the rakes. Indeed in a striking passage in his *History of the Royal Society*, Sprat appealed to these, claiming that in the new science they could enjoy 'the *pleasures* of their *senses* . . . without guilt or remors'. The Royal Society included among its Fellows men who were criticised for their immoral behaviour, such as Lord John Vaughan, called by Samuel Pepys 'one of the lewdest fellows of the age' and the dedicatee of Dryden's scandalous play *Limberham*, although he had serious scientific interests and became

[26] Cudworth,[314] I, 212.
[27] See above, pp. 138, 152; Cudworth,[314] II, 608; cf. Cudworth to Boyle, 16 Oct. 1684 Boyle,[71] VI, 511; Edwards, *Some Thoughts*, pp. 85–6.

President of the Royal Society in the 1680s.[28] It is hardly surprising that the conservatives whose opinions were considered in the last chapter were inclined to conflate the new science with the sceptical values of the rakes. In a sermon at Westminster Abbey in 1667, Robert South alluded to 'a set of fellows got together and formed into a kind of diabolical society, for the finding out new experiments of vice . . . obliging posterity with unheard of inventions and discoveries in sin; resolving herein to admit of no other measure of good and evil, but the judgment of sensuality'. Henry Stubbe took a similar line, referring to the virtuosi as 'our *Comical wits*', whose '*extravagant* opinions' seemed to him to have contributed to 'the growth of *Atheism* in this *Age*'.[29]

Their views might seem confirmed by the co-called 'Ballad of Gresham Colledge', written 'In Praise of that choice Company of Witts and Philosophers who meet on Wednesdays weekly att Gresham Colledg', probably in 1662. These doggerel stanzas were taken by some as a manifesto of at least part of the Royal Society, and they reveal an iconoclastic, bantering and potentially dangerous aspect of the new science. Sentiments like 'Aristotle's an Asse to Epicurus' showed clear links with the values that the orthodox mistrusted.[30] Scholars who have believed the Ballad to be by a Fellow of the Royal Society, however, have been mistaken: in fact it was the work of a hanger-on of the new science, William Glanville, John Evelyn's brother-in-law, with whom (perhaps significantly) Evelyn was later to fall out for his Deistic tendencies.[31]

Part of science's problem was the way in which the genuinely committed merged into the idly curious, those fashionable and mindless fellow-travellers whose proclivity to magisterial pronouncements on topics about which they knew little was criticised. Just as scholars complained how 'the young gallants of the Town' condemned Aristotelian learning 'only because they heard it censurd in the last Coffe house, or think it out of vogue in the Royal Society', so it was felt that the new philosophy in a derivative and superficial form fed the sceptical mood of the day: 'by it they have given the

---

[28] Sprat,[92] p. 344; Pepys,[163] VIII, 532–3; Wilson,[288] passim; Wood, *Life and times*, II, 559; but see Vaughan to Oldenburg, 12 Aug. 1675, 1 Aug. 1676, Oldenburg,[121] XI, 455, EL V, fol. 31, Vaughan to Evelyn, 1 Aug. 1676, Add. 15858, fol. 179.

[29] South, *Sermons*, I, 375; Stubbe, *P.U.*, p. 36. Cf. Stubbe, *Censure*, pp. 55f., *Legends*, sigs. * 3v–4, A3. [30] Stimson,[286] p. 109 and passim; Hull,[233] II, 323–4.

[31] Stimson,[286] p. 104; Cope,[137] p. 12 and n.40; see also Nicolson,[148] p. 115 n.20; Robinson to Power, 30 Oct. 1662, Sloane 1326, fol. 104v; Evelyn,[164] V, 497–8.

ignorant Atheists (for so are most of that perswasion) some plausible pretences for their incredulity without any real ground'. Likewise, Sprat's *History* was cited by the heterodox Charles Blount, and science certainly abetted 'atheism', if through a misuse about which serious virtuosi could do little.[32]

More irksome, the most notorious heterodox philosophers were themselves interested in science, and only an anachronistic definition could exclude Hobbes and Spinoza from the title of 'scientist'. Apart from his great enthusiasm for mathematics, Hobbes had researched on topics like optics which were of general interest in the scientific community. He was also friendly with many men of science and respectful of the Royal Society. He claimed (in a manner shocking to conservatives) that 'As for natural philosophy, is it not removed from Oxford and Cambridge to Gresham College in London' – by which (as Aubrey explained) he meant 'the Royall Societie that meetes there' – 'and to be learned out of their gazettes?'[33]

Spinoza also had scientific interests and he first communicated with English intellectuals in a lengthy correspondence with Henry Oldenburg and, through him, with Robert Boyle. Throughout the 1660s their letters discussed all sorts of scientific ideas and methods, though a break occurred after the publication of Spinoza's *Tractatus Theologico-Politicus* (1670), which apparently embarrassed Boyle and Oldenburg. Contact was re-established only in 1675, when Oldenburg explained how he had at first found the *Tractatus* irreligious due to accepting 'the standard set by the crowd of theologians' rather than considering it on its own merits (understandably enough in the outcry caused by the book's appearance). But he urged Spinoza not to put anything in his next work 'which may seem to undermine in any way the practice of religious virtue, especially since this degenerate and wicked age seeks nothing more eagerly than that sort of doctrine whose conclusions appear to encourage the most riotous vices'.[34]

This was, perhaps, skating on thin ice, and such intercourse was a

---

[32] Bohun to Evelyn, 1668, Evelyn, *Corr.*, no. 301; cf. [Richard Graham, Viscount Preston], *Angliae Speculum Morale* (London, 1670), pp. 44–5; John Keill, *An Examination of Dr. Burnet's Theory of the Earth* (Oxford, 1698), p. 19; Redwood,[279] p. 495; cf. Colie,[277] p. 30 n. 30.

[33] Frithiof Brandt, *Thomas Hobbes' mechanical conception of nature* (Copenhagen, 1928); British Library MS Harley 3360; Hobbes,[70] VI, 348; Aubrey,[161] I, 372.

[34] Colie,[278] pp. 194–5; Oldenburg,[121] I–II, passim; Oldenburg to Spinoza, 8 June, 22 July 1675, ibid, XI, 340, 415.

tangible and embarrassing reminder that Hobbes and Spinoza merely took to extremes the challenge to traditional philosophical arguments and intellectual priorites of the new science as a whole. They arguably drew out implications of the new philosophy which others shied at, and there was no obvious dividing line between thinkers commonly considered pernicious and other devotees of the new philosophy who thought it perfectly harmless. The dilemma was illustrated most clearly by Descartes. The similarity between his world-view and Hobbes's was noted in chapter 1, and there was hence a danger that his mechanistic principles might lead in the same atheistic direction. Yet, as A. N. Whitehead put it, for science and the modern world the mechanism of which Descartes was the most persuasive exponent was 'a general idea which it could neither live with nor live without'.[35]

There can be no doubt of the strong Cartesian influence on many seventeenth-century English scientists. Descartes's mechanical principles of scientific explanation were undeniably intellectually exhilarating, as was their application to an ever wider range of phenomena. Among the most fruitful achievements of the Interregnum and Restoration was Boyle's remodelling into his corpuscular theory of matter the atomist ideas pioneered by the ancient Greeks and revived by Descartes and other thinkers like Pierre Gassendi. No less characteristic were attempts to devise a mechanistic theory of the creation of the world and such cataclysms as the Flood, which were taken quite seriously at the time though often scorned since. The most famous (and most Cartesian) of these was Thomas Burnet's in his *Sacred Theory of the Earth*, the first volume of which appeared in 1681, which was widely and not unappreciatively discussed in scientific circles.[36] Others who sought purely physical explanations of the beginning and end of the world included Edmond Halley.[37]

Yet such projects aroused widespread reservations. Atomism had strong atheistic connections, not least because it had been espoused in antiquity by Epicurus, whose irreligious opinions were quite open and whose questionable moral attitudes were thought naturally to stem from this. The presbyterian divine, Richard Baxter, was typical in conflating modern with ancient materialists in his *Reasons of the*

[35] Whitehead, *Science and the modern world* (Cambridge, 1926), p. 74.
[36] Kubrin,[290] ch. 5; Birch,[118] IV, 69, 83; Newton to Burnet, Jan. 1681, Newton,[127] II, 329–34; JBO VIII, fols. 235–6; Yonge to Hooke, 6 Jan. 1685, EL XYZ, fol. 1; Add. 10039, fols. 65–73. [37] Kubrin,[290] ch. 9; Schaffer,[291] esp. pp. 27–9.

Christian Religion (1667) and attacking 'the *Epicurean* (or *Cartesian*) Hypothesis'. He complained that theorists who tried to reduce everything to matter and motion were 'like idle Boys, who tear out all the hard leaves of their books, and say they have learn'd all when they have learnt the rest'. The anxiety that this attack caused in scientific circles is shown by John Beale's view – passed on by Oldenburg to Boyle – that Baxter should be refuted, though nothing came of this.[38] It was the section expounding atomist views in William Petty's *Discourse . . . Concerning the Use of Duplicate Proportion* (1674) which aroused an orthodox cleric to write a letter in reply to it, apart from what he saw as the work's metaphysical solecisms. 'I never yet read of any *Anthems* composed for the Contemplation of Atoms', wrote another critic, and it is easy to understand the anxiety of Cudworth and others to demonstrate that true, as against corrupt, atomism could be traced back to Moses.[36]

The reactions to Burnet and other 'world-makers' were no less extreme, for mechanical theories were seen to detract from God's personal activity in the creation of the world and hence to strengthen atheist arguments: *The Sacred Theory of the Earth* sparked off an extensive controversy, while Halley also won a dangerous reputation for his mechanistic views from which he had to vindicate himself.[40] The first and perhaps most representative expression of orthodox reservations about Burnet was penned by Herbert Croft, Bishop of Hereford, who protested the dangers of explaining the creation by mechanistic principles, preferring to affirm the inscrutable activity of God's providence. 'This way of Philosophising all from Natural Causes, I fear, will make the whole World turn Scoffers': Burnet should 'lay aside his curious vain and endless Philosophising labour to find out the Natural Causes thereof', for the testimony of scripture was 'far more considerable than his trivial experiments of a Mathematical Instrument or Cubical Pot'.[41]

Some went on from such misgivings to label the new science as a

---

[38] Baxter, *The Reasons of the Christian Religion* (London, 1667), p. 498; Oldenburg to Boyle, 24 Dec. 1667, Oldenburg,[121] iv, 80.

[39] Thomas Barlow, *The Genuine Remains* (London, 1693), pp. 151, 153–5; Thomas Manningham, *Two Discourses* (London, 1681), p. 95; see above, p. 150.

[40] Michael Macklem, *The anatomy of the world* (Minneapolis, 1958), ch. 3 and appendix 1; Nicolson,[330] pp. 233f; Levine,[261] pp. 25f; Schaffer,[291] p. 17 and passim; Kubrin,[290] pp. 246f; Halley,[238] pp. 266–7.

[41] Croft, *Some Animadversions Upon a Book Intituled The Theory of the Earth* (London, 1685), pp. 40–1, 99.

whole as 'very apt to be abused and to degenerate into *Atheism*' by encouraging a neglect of spiritual knowledge. 'I had rather consider the *Rain-bow* as the *Reflection* of God's Mercy, then the Sun's Light', wrote Thomas Manningham, a young divine, attacking natural philosophy as 'the great *Diana* of this *Mechanick* age'. 'Is there anything more Absurd and Impertinent', John Norris of Bemerton scornfully asked, than to find 'a Man, who has so great a Concern upon his Hands as the Preparing for Eternity, all busy and taken up with *Quadrants*, and *Telescopes, Furnaces, Syphons* and *Air-Pumps?*'[42] In the 1670s things had reached such a pass that Robert Boyle complimented the Oxford divine, Thomas Smith, on refraining from such strictures in his recent sermon on 'the Credibility of the Mysteries of the Christian Religion': 'the virtuosi will I hope easily take notice of your discreet moderation in declining the opportunity that would have mislead a less ingenious person to fall foul upon Philosophy, which many thinke it, tho erroneously enough, a great service to Christianity to decry'.[43]

That there was an element of truth in such charges was, however, admitted by scientists themselves. Sprat agreed that 'some *Philosophers*, by their carelesness of a Future Estate, have brought a discredit on *Knowledge* itself', even though arguing that 'this rather proceeds from their own *Genius*, than from any corruption that could be contracted from these *Studies*'. Similarly Boyle justified his *The Excellency of Theology, Compared with Natural Philosophy* (1674) on the grounds 'that under-valuation . . . of the study of things sacred . . . is grown so rife among many (otherwise ingenious) persons, especially studiers of physicks, that I wish the ensuing discourse were much less seasonable than I fear it is'. Anxiety about the possible threat of their studies to orthodoxy partly explains the inordinate amount of time that many eminent naturalists spent in wrestling with the problem of science's relations with religion, even though none of them found on examination that such worries were justified. There is an element of truth in R. S. Westfall's view that 'more than answering hypothetical atheists, they were trying to satisfy their own doubts', indeed that 'the virtuosi nourished the atheists within their own minds'.[44]

---

[42] Casaubon,[265] p. 30; Manningham, *Two Discourses*, pp. 2, 98; Norris, *Reflections upon the Conduct of Human Life* (London, 1690), p. 132.
[43] Boyle to Smith, Dec. 1675, Smith 48, fols. 5–6.
[44] Sprat,[92] p. 375; Boyle,[71] IV, 2; Westfall,[295] pp. 145, 219 and passim.

Most scientists not only reassured themselves about this, but felt a genuine sense of providence in scientific discovery, which they sometimes recorded. Having resolved the problem for themselves, however, they then turned to vindicate their studies for all, producing a large literature protesting the religious benefits of their enterprise and illustrating how scientists 'have raised *impregnable Ramparts* with much *industry* and *pious pains* against the *Atheists*'. 'Though I am willing to grant', Boyle wrote, 'that some impressions of God's wisdom are so conspicuous, that . . . even a superficial philosopher may thence infer, that the author of such works must be a wise agent; yet how wise an agent he has in those works expressed himself to be, none but an experimental philosopher can well discern'.[45]

The classic statement of this position was John Ray's *The Wisdom of God Manifested in the Works of the Creation*, published in 1691 but based on lectures that Ray had given in Cambridge in the 1650s. Its arguments owed something to earlier writings of Henry More, but in Ray's hands they blossomed, and it is not surprising that this proved Ray's 'most popular and influential achievement', going into edition after edition. 'There is no greater, at least no more palpable and convincing Argument of the Existence of a Deity', he wrote, 'than the admirable Art and Wisdom that discovers itself in the make and constitution, the order and disposition, the ends and uses of all the parts and members of this stately fabrick of Heaven and Earth.'[46]

'What variety? What beauty and elegancy? What constancy in their temper and consistency, in their Figures and Colours?', Ray soliloquized about stones. It was even thoughtful of God to allow the rain to fall gently rather than all at once, while his teleological enthusiasm came to a climax in praising the human body: 'there is nothing in it deficient, nothing superfluous, nothing but hath its End and Use'.[47] Ray argued that the study of God's handiwork was not just a luxury but a positive Christian duty, and both he and Boyle espoused a view of 'the Scientist as Priest'. This ideology spread widely, not least in the provinces, where virtuosi espoused 'the great and worthy designe of an Universall History of Nature to be borne

---

[45] Flamsteed,[162] p. xxii; see also Hooke,[65] p. xxviii; Glanvill,[66] p. 138; Boyle,[71] v, 517.
[46] Raven,[135] p. 452 and ch. 17 passim; Ray, *The Wisdom of God Manifested in the Works of the Creation* (London, 1691), pp. 11–12.
[47] Ibid, pp. 67, 65, 155.

out of her owne bowells by faithfull experiments for the confusion of growing Atheisme, and Philosophy falsely soe called'.[48]

Thus presented as a pious and industrious enterprise, it was easier to deny any taint in science of the atheistic tendencies of fashionable circles, and its protagonists stressed the contrast between the two. William Petty deliberately distinguished the aims of science '(both as to *difficulty* and *dignity*) from what is commonly called Wit (and which takes with far the greater part of Mankind)', and it was commonly felt that the study of nature was more worthwhile than trivial literary compositions. This echoed Boyle's Interregnum ideas about the acquisition of useful knowledge as an ethical undertaking in contrast to the indolence of 'Gallants' and 'Profane Persons'. Though one modern author has linked science with liberalising values and a lessening of religious pressure, he neglected the extent to which scientists aligned themselves with 'asceticism'.[49]

The antithesis between science and worthless fashionable pursuits can only have been strengthened in the minds of the virtuosi when their enterprise was subjected to ridicule and to more practical obstacles. Attempts were made to sabotage the Royal Society's blood-transfusion experiments and to deposit the characteristic 'Ballets and Boufonries' of 'these scoffing times' in the Society's archives.[50] Satire may have partaken to some extent of the humanist critique of science referred to in the last chapter, but to its victims it looked more like a negative emanation of 'wit', out to raise an easy laugh by any means possible. Indeed, after seeing some 'Scurrilous Interlude' in 1676 (probably Shadwell's *The Virtuoso*, the chief work of the genre), John Evelyn went home to write a letter to his confidante, the saintly but suggestible Margaret Blagge (later Mrs Godolphin). In it he lectured her in detail on the achievements of the new science in 'attaining to the knowledge of the Aspectable works of God', and warned her not to be 'prejudic'd at the Scoffs & Raillery, of the Bouffoones, & ignorant Fops of this abandond age . . . who to the detriment of their Soules & Bodies consume their precious moments either in pursuite of their vices, or insignificant or

---

[48] Ray, *The Wisdom of God*, pp. 122f; Harold Fisch, 'The scientist as priest: a note on Robert Boyle's natural theology', *Isis*, 44 (1953), 252–65; Cole to Plot, 8 Dec. 1684, Gunther,[129] p. 244.

[49] Petty, *The Discourse . . . Concerning the Use of Duplicate Proportion* (London 1674), sig. A6; Birch,[118] I, 27; Jacob,[82] p. 6 and passim; Feuer,[209] chs. 1–2.

[50] Skippon to Ray, [Dec. 1667], Ray,[124] p. 23; Oldenburg to Boyle, 28 Jan. 1668, Oldenburg,[121] IV, 121.

impertinent diversions'. In 1679 Evelyn made public very similar sentiments in a long and plaintive addition to the preface to the third edition of *Sylva*.[51]

More significant were the relations between the Royal Society and Thomas Hobbes, for these apparently reveal disagreement among scientists and an attempt to establish an orthodox centre for the new science in open opposition to that unconventional philosopher. A recent revaluation has shown that (contrary to common belief) Hobbes was certainly not 'excluded' from the Royal Society as if it were a modern learned society: in fact, as shown in chapter 2, membership and non-membership was often quite haphazard. If Hobbes was not encouraged to join, this was not least because his argumentative manner meant that the Society did not want to be saddled with a club bore. Indeed Hobbes had many admirers among the Fellows, including John Aubrey, who presented the Society with a portrait of the philosopher which was hung in its rooms, while the Society actually approached Hobbes in 1675 to see if he had any material that they might publish.[52]

But – though perfectly true – this is only half the story, for Hobbes's friends in the Society were balanced by enemies. Thus in 1673 Sir John Hoskyns, a Fellow favourable to Hobbes, complained to the like-minded Aubrey that the Secretary, Oldenburg, 'with some regret, grumbles out an account or notice rather, of your Country-man Thomas Hobbes his last booke'. He went on to regret Oldenburg's treatment of 'that stout old blade that defyes limits of enquiry. You know what influence is against T. H.': 'They must shut a roome very close that will have no species admitted but thorough their owne glasse', he reflected. Aubrey was inclined to attribute such attitudes to personal antagonism, but others were convinced that the room should be shut just so close that experimental philosophy should not seem distasteful to the orthodox whom Hobbes displeased.[53] Hence prominent Fellows did not just ignore Hobbes but took the offensive, and Robert Boyle and John Wallis wrote books attacking Hobbes's science in the hope of discrediting both that and his whole method of reasoning.

Boyle was most explicit in his *Examen of Mr T. Hobbes his Dialogus Physicus De Natura Aeris* (1662), an experimental refuta-

[51] Evelyn to M. Blagge, 18 July 1676, Evelyn, *Lb.*, no. 382; Evelyn,[197] sigs. A1–4, a1.
[52] Skinner,[109] pp. 218, 238 and passim; Aubrey,[161] I, 354–5, 372; see above, p. 47.
[53] Hoskyns to Aubrey, 22 Nov. 1673, Aubrey 12, fol. 213; Aubrey,[161] I, 372.

tion of Hobbes's theory of the nature of air which defied retaliation. 'It was also suggested to me', he wrote, .

that the dangerous opinions about some important, if not fundamental, articles of religion, I had met with in his *Leviathan*, and some other of his writings, having made but too great impressions upon divers persons (who, though said to be for the most part either of greater quality, or of greater wit than learning, do yet divers of them deserve better principles) these errors being chiefly recommended by the opinion they had of Mr *Hobbes*'s demonstrative way of philosophy; it might possibly prove some service to higher truths than those in controversy between him and me, to shew, that in the Physics themselves, his opinions, and even his ratiocinations, have no such great advantage over those, of some orthodox Christian Naturalists.[54]

Wallis assailed Hobbes at an even more fundamental level. Knowing that Hobbes saw mathematics as the keystone of his whole system of thought, he sought to 'show him by the reverse process of reasoning, how little he understands the Mathematics from which he takes his courage'. He laid bare Hobbes's shortcomings as a geometer 'the rather because he hoped that the discovery of them would lessen Mr Hobbes's credit in his other writings, which he was sensible was at that time too great, considering their influence'.[55]

Indeed Hobbes was seen as the extreme exponent of dogmatism and attacked as such. Wallis and Boyle disliked him as 'one highly *Opinionative and Magisterial*', a man so intellectually arrogant that he believed he could square the circle. In contrast, they stressed the advantages of Baconian experiment, and similar charges were levelled at other unorthodox thinkers, not least Spinoza, in his letters to whom Oldenburg sang the praises of inductive inquiry as a surer way to scientific truth than the *a priori* reasoning of ancient and modern system-builders. The fear of atheism thus added urgency to the championship of Baconianism and the hostility towards premature systematisation referred to in chapter 1. Similarly, in attempts to confute atheistic principles the speculative and dogmatic nature of irreligious ideas were contrasted with the reliable if incomplete deductions of Baconian science.[56]

---

[54] Boyle,[71] I, 187.
[55] Wallis to Huygens, 1 Jan. 1659, translated in J. F. Scott, *The Mathematical Work of John Wallis* (London, 1938), pp. 170–1; ibid, p. 170 and ch. 10 passim. For a slightly different view of Boyle's and Wallis's relations with Hobbes see Skinner,[109] pp. 220–2, 224–5, 227–8.
[56] Wallis, *Hobbius Heauton-timorumenos*, p. 3; Hobbes, *Quadratura Circuli* (London, 1669); Boyle,[71] IV, 105–6; see also Thomas Sprat, *Observations on Monsieur*

For atheism was not infrequently seen as a form of dogmatism rather than (as might have been expected) an extreme scepticism. This explains Henry More's conviction of 'the great Affinity and Correspondency betwixt *Enthusiasm* and *Atheism*', which might otherwise seem like an empty smear. It explains, too, how scholasticism – almost proverbially dogmatic – could appear a breeding ground for infidelity, while similar aspersions were cast on Roman Catholicism. There is even something to be said for R. F. Jones's perceptive view of a polarisation between 'empiricism' and 'rationalism' in Restoration thought and writing, with Baconian scientists championing the former against what they saw as the dangerous logical corollaries of the latter.[57]

This may also be the context in which to see the establishment of a new methodological orthodoxy among scientists, a growing distrust of philosophical certainty and a preparedness to be content with probabilities. In this, scientists moved away not only from thinkers like Hobbes but even from Bacon (and, for that matter, from Aristotle). 'Probabilist' ideas seem to have first been put forward by Marin Mersenne and Pierre Gassendi, two French natural philosophers who much influenced English thought in the 1650s, and it is among Interregnum scientists that this viewpoint is first found, to be developed more fully after 1660.[58] After being expressed by Boyle it was later elaborated by John Locke in an epistemological statement that has sometimes been seen as characteristic of Restoration science and was full of implications for the future.[59]

Method, however, was only the beginning. It was more important to use it and to prove God's activity in the world, which required more than pious commonplaces about the exquisite construction of nature like those in Ray's *The Wisdom of God*. It was vital to refute those who saw the world as a self-contained machine, whether dead (like Hobbes) or alive (like Spinoza), and the danger of atheism overrode the generally tolerant nature of the science of the time,

de Sorbier's *Voyage into England* (London, 1665), pp. 232–5; Hall,[32] p. 242 and passim; Bentley,[316] III, 27–200 passim.

[57] More, 'Enthusiasmus Triumphatus' in *A Collection of Several Philosophical Writings* (London, 1662), p. 1 (sig. s4); cf. Manningham, *Two Discourses*, pp. 7–12; Jacob,[301] pp. 217–18; Jones,[241] pp. 110–11.

[58] Wood,[28] pp. 260f; Aarsleff in Gillispie,[134] XIV, 367; Teague,[27] pp. 223–6.

[59] Yolton,[38] pp. 114–17 and passim. See also Van Leenwen,[86] chs. 3–5, G. A. J. Rogers, 'Boyle, Locke and reason', *J.H.I.*, 27 (1966), 205–16, and M. J. Osler, 'John Locke and the changing ideal of scientific knowledge', ibid, 31 (1970), 1–16.

putting a premium on the establishment of a 'safe' orthodoxy. Even Cartesianism raised difficulties, as John Worthington recognised, complaining to Henry More how Cambridge students were 'enravisht' with it

and derive from thence notions of ill consequence to religion . . . And seeing they will never return to the old Philosophy, in fashion when we were young scholars, there will be no way to take them of from idolizing the French Philosophy, and hurting themselves and others by some principles there, but by putting into their hands another Body of Natural Philosophy, which is like to be the most effectual antidote.[60]

Indeed he was not alone in seeing Descartes as hardly less pernicious than Hobbes, whatever others thought of the safety of Descartes's dualist separation of mind and matter in leaving the spiritual realm intact.

Of the world-views that resulted, that of the Cambridge Platonists most obviously reflected this dilemma. Henry More had at first welcomed Descartes's account of the mechanical philosophy, in part as an answer to Interregnum 'enthusiasts', though he was always aware that such a system was important not least for what its rigid devotion to matter in motion left unexplained. But More became increasingly alarmed by the pernicious implications of mechanistic world-views, particularly vulgarised Cartesianism which was not offset by its originator's piety. In the 1650s he therefore worked out a rival cosmology, trying to answer the atheist threat by producing his own amalgam of Platonic and Cartesian ideas in which a prominent role was played by a '*Spirit of Nature*', a concept which he expounded in his philosophical writings both before and after 1660. In the Restoration he even tried to use Boyle's findings to bolster his hypothesis of an '*Immaterial* Being that exercises its *directive* Activity on the *Matter* of the World'.[61]

Ralph Cudworth, too, was anxious to reinstate vitalism in the universe to purge the dangers of materialism. In his *True Intellectual System of the Universe* (1678) he argued that

since neither all things are produced fortuitously, or by the unguided mechanism of matter, nor God himself may reasonably be thought to do all

---

[60] Worthington to More, 29 Nov. 1667, in *The diary and correspondence of John Worthington*, ed. R. C. Christie, II part 2 (Manchester, 1886), 254.

[61] Gabbey[298]; Pacchi,[297] chs. 1, 4; More, 'An Antidote Against Atheism' in *A Collection of Several Philosophical Writings*, pp. 43–6 (sigs. E1–2); R. A. Greene, 'Henry More and Robert Boyle on the spirit of nature', *J.H.I.*, 23 (1962), 451–74.

things immediately and miraculously; it may well be concluded, that there is a plastic nature under him, which, as an inferior and subordinate instrument, doth drudgingly execute that part of his providence, which consists in the regular and orderly motion of matter.

Writers influenced by the Cambridge Platonists who took refuge in the doctrine of spirits to avoid the dangerous implications of Cartesian dualism included Henry Power, John Ray and Joseph Glanvill.[62]

In Glanvill's case this led to an obsessive interest in cases of witchcraft, which he collected and published with the encouragement of More and others to confound Hobbes, Spinoza and other 'such course-grain'd Philosophers'. As More explained in a letter that accompanied the work: 'I look upon it as a special piece of Providence that there are ever and anon such fresh Examples of Apparitions and Witchcrafts as may rub up and awaken their benummed and lethargick Minds into a suspicion at least, if not assurance, that there are other intelligent Beings besides those that are clad in heavy Earth or Clay.' Boyle also supported Glanvill's witchcraft project, and he agreed with the Cambridge Platonists in stressing the feasibility of miraculous intervention in the world. The possibility of miracles was also at the heart of the argument against Spinoza that he pursued in a lengthy manuscript tract.[63]

Though their preoccupations overlapped, however, Boyle parted company with the Cambridge Platonists in his view of the workings of the world. He was as aware as Cudworth of the dangers of Hobbesian materialism, but he also saw the risk that occult principles like Cudworth's and More's could be taken as natural: God would thus be identified with nature and the possibility of his action outside it implicitly denied, so that his power would be reduced as much as by Hobbesian materialism. In a sense he was right, and the Cambridge Platonists' unwitting secularising tendency has also been noted in morality.[64] Boyle was equally hostile to any view which seemed to imply that matter was purposive, including scholastic science and what has recently been called 'pagan naturalism', those

---

62 Cudworth,[314] I, 223–4; Colie,[296] ch. 7; Mintz,[274] ch. 5; Webster,[29] pp. 173–4; Ray, *The Wisdom of God*, pp. 20f.

63 Glanvill,[315] p. 26; Prior,[299] pp. 169, 183–4 and passim; Cope,[137] chs. 4, 6; Colie,[278] pp. 198–9, 211–19.

64 McGuire,[302] pp. 534–5, 542; Westfall,[295] p. 84; J. A. Passmore, *Ralph Cudworth: an interpretation* (Cambridge, 1951), pp. 83–4 and passim.

pantheistic ideas that Cudworth saw as a subsidiary threat.[65] Instead, Boyle stressed the inertness of matter and God's direct control of it, so that all the workings of nature were direct manifestations of the omniscience and goodness of a constant and intelligent power: this reflected the strong voluntarist streak in Boyle's thought.[66]

More important still as a solution to the materialist dilemma was Newton's blend of elements of the mechanistic tradition with the more vitalist streak of the Cambridge Platonists. Reacting against Cartesianism, he tried to modify mechanism in an acceptable way without resorting to the Platonists' occultism. His answer was to domesticate the concept of force as a quantifiable, non-mechanical element without which any world-view was incomplete, while following Boyle in postulating the universe as passive matter directed by providence. Newton's God was as untrammelled as Boyle's, a being who (in the words of the 'General Scholium' to the *Principia* of 1713) 'governs all things, not as the soul of the world, but as Lord over all'.[67] Views differed on how Newton's 'force' was to be interpreted, some seeing it in a more occult light than others, and Newton himself vacillated on related points.[68] But it was widely held to give the coup de grâce to thorough-going materialism and hence to provide the 'safe' system of the world according to the prevailing mechanistic principles that the age required. As Richard Bentley put it: 'we have great reason to affirm, that universal gravitation, a thing certainly existent in nature, is above all mechanism and material causes, and proceeds from a higher principle, a divine energy and impression'.[69]

Furthermore, as has often been noted, Newton's principles were deliberately popularised as a satisfying world-view vindicating God's control over the natural order and hence proving the error of materialist doctrines and, by extension, the immoral life style that they were seen to accompany.[70] The agency of this was a lectureship set up under a provision in the Will of Robert Boyle, who died in

[65] Jacob,[300] esp. pp. 275–7; Jacob[301]; Cudworth,[314] I, xl–xli, 215–16. See above, p. 169.
[66] McGuire,[302] esp. pp. 535f; Burtt,[303] ch. 6. See also F. Oakley, 'Christian theology and the Newtonian science: the rise of the concept of the laws of nature', *Church History*, 30 (1961), 433–57.
[67] Westfall,[47] ch. 7; Newton,[63] p. 544; see also Burtt,[303] ch. 7, esp. sect. 6.
[68] Kargon,[50] pp. 136–9; J. E. McGuire, 'Force, active principles, and Newton's invisible realm', *Ambix*, 15 (1968), 154–208; Westfall,[47] pp. 391f.
[69] Bentley,[316] III, 165.
[70] Hessen,[167] pp. 185–6; Redwood,[272] ch. 4; and see the studies cited on pp. 215–16 below.

1691, at which eminent divines gave sermons to a large London audience 'for proving the Christian Religion, against notorious Infidels, *viz.* Atheists, Theists, Pagans, Jews, and Mahometans, not descending lower to any Controversies, that are among Christians themselves' (from the first the lectures tended to concentrate on the first category of infidels, paying less attention to Jews and Mahometans). The popular nature of the project was clear: care was taken that as many people as possible heard the lectures, and even the time of year and day of the week of the series were changed to this end.[71]

The first set of lectures was delivered in 1692 by the young scholar and divine, Richard Bentley, whose views on the nature of the atheist threat have already been noted. Published in 1694 with the title *A Confutation of Atheism*, the lectures expounded the customary argument from design, buttressed by the findings of the new science. Much of Bentley's text went through unbelievers' claims and rebutted them, using discoveries by scientists like Boyle, Malpighi and van Leeuwenhoek, but in his seventh and eighth lectures he diverged more than elsewhere from his refutation of infidel notions to give an exposition of Newton's cosmological ideas, and Newton figured more than any other scientist in his notes.[72] By using the *Principia* Bentley was thus able to add a superior cosmological dimension to the traditional teleological picture, asserting that the perfection that he described in the universe should 'be ascribed to the transcendent wisdom and benignity of that God, *who hath made all things very good, and loveth all things that he hath made*'.[73]

Newton was delighted that his cosmology was thus presented in relatively simple terms to vindicate God's active role in the world. He gladly helped Bentley to avoid solecisms in the printed version of his sermons, confirming and elaborating the younger man's interpretation of his ideas to prove the existence of an intelligent being in the world. 'When I wrote my treatise about our Systeme' he explained, 'I had an eye upon such Principles as might work with considering men for the beliefe of a Deity & nothing can rejoyce me more then to find it usefull for that purpose.'[74] He had initially considered presenting

---

[71] Jacob,[307] pp. 144, 148–9, 159–60; Evelyn to Tenison, 23 Feb. 1694, Evelyn, *Lb.*, no. 691.   [72] Bentley,[316] III, 27–200 passim.   [73] Bentley,[316] III, 185.
[74] Newton to Bentley, 10 Dec. 1692, Newton,[127] III, 233. Cf. also Newton to Bentley, 17 Jan., 11, 25 Feb. 1693, ibid, III, 238–40, 244, 253–6, Bentley to Newton, 18 Feb. 1693, ibid, III, 246–52.

his *Principia* in a popular format, and had advocated a public exposition of the design of the universe: it is even possible that he was a party to the Boyle Lectures' foundation. Newton's preoccupation with materialism and its dangers is further illustrated by the famous exchange in 1693 when he accused John Locke of being 'a Hobbist' and claimed that his *Essay Concerning Human Understanding* struck at the roots of morality.[75]

Bentley's use of Newtonian argument was echoed in later Boyle Lectures and more widely. It has thus been claimed that a self-conscious new ideology triumphed in the early eighteenth century, in which Newton's system of the universe was expounded to guard the theological consensus against the attacks of extremists. A parallel development has been traced in the writings on the origins and future of the world referred to earlier. Here, too, Newton's disciples tried to guard against atheism by arguing the impossibility of a purely mechanical theory either of the creation or destruction of the world, so that (as with gravity in the universe) the immediate activity of God was essential. Bentley attacked Thomas Burnet in his Boyle lectures, while – with Newton's approval – the mathematician, John Keill, condemned all such mechanistic speculations as premature and dangerous, attempting a scientific demonstration of their impossibility. 'These flood-makers have given the Atheists an argument to uphold their cause, which I think can only be truely answer'd by proving an universal Deluge from Mechanical causes altogether impossible', he wrote: 'no secondary causes without the interposition of Omnipotence could have brought such an effect to pass'.[76]

But, though there was undoubtedly a 'Newtonian' lobby in the earth sciences, it is an exaggeration to postulate any unanimity in such matters, since opinion remained inchoate. Even in astronomy, not all accepted Newton's view of gravity, like Sir Christopher Wren, who 'smiles at Mr Newton's belief that it does not occur by mechanical means, but was introduced originally by the Creator'. It has been argued, moreover, that early eighteenth-century Newtonianism was far from homogeneous.[77] Cartesian ideas remained common despite efforts by Newtonian popularisers: an English transla-

---

[75] Newton,[63] p. 397; Jacob,[307] pp. 153–5; Newton to Locke, 16 Sept. 1693, Newton,[127] III, 280; Manuel,[139] p. 216.

[76] Jacob[307]; Bentley,[316] III, esp. 193f; Kubrin,[290] pp. 219f, 319f. and chs. 10–11 passim; Keill, *An Examination of Dr Burnet's Theory of the Earth*, pp. 21, 178–9.

[77] Porter,[46] pp. 77–8 and ch. 3 passim; Roy Porter, 'Gentlemen and geology', *Histori-*

tion of Anthony Le Grand's popular exposition of Descartes was published with extensive patronage from members of the fashionable establishment in 1694, the same year as Bentley's Boyle Lectures, while Fontenelle's *Dialogues on the Plurality of Worlds*, which vulgarised Cartesian astronomical views, were reprinted in the 1690s and frequently in the early eighteenth century.[78] Even in the Boyle Lectures, the more traditional design argument of Ray with its stress on biological vitalism remained as influential and arguably as important as Newtonianism in buttressing the theological consensus.[79] Moreover by no means all the Boyle Lectures had any scientific component at all.

It is even wrong to take it for granted that Newtonianism easily became 'orthodox'. The difficulties involved are shown by the case of Thomas Burnet, that world-maker whom the Newtonians attacked. For Burnet had believed his demonstration of the mechanics of the creation to strike at the view – commonly associated with atheists – that the world was eternal: this threat to the need for a creator preoccupied many, including Newton.[80] Indeed John Evelyn asked Samuel Pepys, on seeing *The Sacred Theory of the Earth*, 'Was ever any thing better sayd to convince the atheist than what he has written concerning matter and motion and the Universal Providence, to the reproch of chance and out contingent fops?'[81] But the dangers of principles like Burnet's were soon discerned by the Newtonians and others who saw that any mechanism must be repeatable and hence could posit a world eternally renewing itself. Burnet thus gained a reputation for heterodoxy which was confirmed when he moved further from accepted views in defence of his theory.[82]

Similar problems faced John Locke, who – despite his attempts to protest his conformity – was widely suspected in the 1690s of being a party to the spread of theological ideas with atheistic implications. His work was condemned by the Middlesex Grand Jury and the

cal Journal, 21 (1978), 813; David Gregory's memorandum, 20 Feb. 1698, in Newton,[127] IV, 266–7; Geoffrey Bowles, 'John Harris and the powers of matter', *Ambix*, 22 (1975), 21–38.

[78] Anthony Le Grand, *An Entire Body of Philosophy, According to the Principles of the Famous Renate Des Cartes* (London, 1694); Layton,[319] pp. 16–18.

[79] See esp. W. Derham, *Physico-Theology: or, a Demonstration of the Being and Attributes of God, from his Works of Creation* (London, 1713).

[80] Kubrin,[290] pp. 91f; Kubrin,[289] esp. pp. 32–9.

[81] Evelyn to Pepys, 8 June 1684, in *Private correspondence and miscellaneous papers of Samuel Pepys*, ed. J. R. Tanner (London, 1926), I, 23–4.

[82] Nicolson,[330] pp. 238–9; Kubrin,[290] pp. 143f.

University of Oxford, and such doctrines as his rejection of innate ideas were commonly seen as potentially irreligious, a suspicion confirmed when Toland and other heretical thinkers adopted his opinions.[83] Though Newton did not suffer Locke's fate in the 1690s, this was arguably because many guardians of theological orthodoxy had not yet absorbed the message of the Boyle Lectures and did not see Newtonianism as important enough to merit attack. By the early eighteenth century, however, they did, and so Newtonians like Samuel Clarke were pilloried for abetting the growth of heretical ideas and their attempts at a scientific defence of religion deprecated.[84] Even in the 1720s the view that 'all Enquiries into Nature' were 'dangerous, and many times prejudicial to Religion' was 'industriously propagated by many Persons'.[85]

If it is mistaken to talk glibly about a Newtonian triumph, however, this is partly because it is misleading to seek a straightforward orthodoxy at all. Though a sense of heterodoxy is much in evidence in these years, the centre of gravity of the conformist party who opposed it is much more difficult to define. In many ways theological argument in this formative period is more easily defined negatively than positively; the fear of atheism was one of the few fixed points of reference, and people arranged themselves by the nature of their reactions to that. More important, perhaps, is the danger illustrated in this chapter of overemphasising the tendencies towards secularism and naturalism in the late seventeenth century. The assault on atheists by scientists represented a strong (if moderate) supernaturalist reaction to these forces. Those theologians who accused the Newtonians of selling the pass by their scientific reasoning, instead stressing the importance of scriptural revelation, prove that there were alternative positions whose vitality is shown by the antagonism that the Newtonians encountered in the eighteenth century.[86]

[83] Yolton,[293] pp. 11, 118f, ch. 2 and passim.
[84] Holmes,[309] pp. 169–70; Stewart;[311] see also Wilde,[310] esp. pp. 6–11.
[85] Benjamin Worster, *A Compendious and Methodical Account of the Principles of Natural Philosophy* (London, 1722), pp. iv–v.
[86] I hope to throw further light on the issues raised in this chapter in a study of atheism in the early modern period that I am currently preparing.

Conclusion

There is a danger of ending a book like this with facile assertions that all that was wanting in the late seventeenth century was realised in the eighteenth, that science almost overnight became orthodox, practical and popular to a degree undreamt of before 1700. A cautionary tale may, however, be sought in the recent historiography of early eighteenth-century politics and political ideology, where simple-minded notions about the triumph of the Whigs have come under scrutiny. It is now clear that Toryism remained vital at a time when Whig dominance had long been assumed, while presumptions that Whiggish ideology was unchallenged also fail to do justice to the continuity of more traditional political argument.[1] England may have witnessed a commercial or a financial revolution in these years, but all spheres of thought and practice were not thereby changed overnight.

This continuing ambivalence is also found in early eighteenth-century science. It is especially clear where science has been brought closest to politics, in arguments that the victory of the Whigs was accompanied by a triumph of Latitudinarians and Newtonians in the church. Some recent scholars may have gone too far in assuming direct links between politics and natural philosophy, but there can be no doubt that the vitality of Toryism in the early eighteenth century reflected the strength of the High Church party, and High Churchmen were far from happy about Newtonianism. These were the pillars of orthodoxy who saw science as dangerously prone to encourage Deism and hence atheism, which seemed no less serious a problem than ever. As prominent a Newtonian as William Whiston, one-time Lucasian Professor of Mathematics at Cambridge, was prosecuted for heresy, and though the failure of Convocation to

[1] See particularly B. W. Hill, *The growth of parliamentary parties 1689–1740* (London, 1976), J. P. Kenyon, *Revolution principles* (Cambridge, 1977).

convict him has been seen as a sign of the times, the episode was nonetheless significant.[2]

Science also remained far from dominant in the intellectual realm surveyed in chapter 6. Though the implicit polarisation between science and learning that Meric Casaubon noted in the 1660s was ultimately to end with science usurping the authority previously held by historical and ecclesiastical scholarship, it is easy to overestimate the speed at which this occurred. The erudition that seemed so important in the late seventeenth century retained its vigour in the early eighteenth in the work of men like Thomas Hearne, Thomas Madox, Edmund Gibson and William Wake. This is a further warning against seeking over-easy correlations between churchmanship and intellectual affairs. For though such scholarship has sometimes been seen to show High Church leanings and disproportionate links with the Non-Jurors, the affiliations of historical and ecclesiastical scholarship remained broad in the early eighteenth century as they had been in the seventeenth.[3]

Continuity is perhaps symbolised most clearly by the Royal Society. The Society's wide esteem throughout Britain and Europe in the late seventeenth century was unabated in the eighteenth. If anything it was increased by a burgeoning of science in the British colonies in America, itself encouraged through correspondence with the Society and buoyed up by the prestige of the Society's name. But the Society's renown still co-existed with tension and decay at its London home. 'Foreigners have just grounds for amazement when they hear how wretchedly all is now ordered', wrote a continental visitor, Zacharias Conrad von Uffenbach, who visited the Society in 1710 with expectations formed by works like the *Philosophical Transactions* and Grew's *Musæum Regalis Societatis* (1681). He was moved to gloomy reflections about the tendency of all public societies to degenerate to a state of 'indifference and sloth'.[4] Personal antagonism remained marked. Though the so-called 'Principate of Newton' – President from 1703 to 1727 – is sometimes acclaimed as a newly decisive era, those excluded from Newton's dictatorial cabal saw a persistence of the destructive strife of the seventeenth century. As

---

[2] Eamon Duffy, ' "Whiston's affair": the trials of a primitive Christian 1709–14', *Journal of Ecclesiastical History*, 27 (1976), 129–50. See above, p. 187.

[3] Douglas,[258] chs. 9–11; ibid, esp. p. 248; G. V. Bennett, *White Kennett 1660–1728* (London, 1957), pp. 247–8.

[4] Stearns,[115] esp. chs. 8–10; W. H. Quarrell and M. Mare, eds., *London in 1710: from the travels of Zacharias Conrad von Uffenbach* (London, 1934), p. 98.

John Woodward commented in 1711 of the rivalry between Newton and Hans Sloane: 'The Royal Society is brought to a very low Ebb and indeed to the utmost Contempt.' Tension between the scholarly and social functions of the Society also remained throughout the eighteenth century, only to be solved by the reforms of the early nineteenth.[5]

There was equally little change in the relations of science with government and economic life surveyed in chapters 4 and 5. If King William's Wars seemed likely to usher in a new era of enlightened use of science in the service of the state, such hopes were soon disappointed. In the early eighteenth century as previously, scepticism predominated in high places about the ambitious claims of scientists. John Flamsteed found the government as parsimonious after 1700 as before, and even in as obvious a sphere for mutual assistance as maritime exploration, progress was disappointing to scientific enthusiasts. Arguably the potential for co-operation in this field was only realised late in the eighteenth century in the voyages of Captain Cook, for which the government paid and the Royal Society and the Admiralty were jointly responsible.[6]

Indeed, in the eighteenth century the contrast between the French government's espousal of science and its independent and ineffectual role in England became more intense than before. As the Académie des Sciences went from strength to strength as the intellectual embodiment of the state, the Royal Society seemed an ever paler reflection. So too in education, for there was no English equivalent to the technical schools pioneered in France at this time.[7] In England, scientific education was left in the eighteenth century as earlier to private initiative. Though it took root to a greater extent after 1700, scepticism has been expressed about its scale and significance even then. Movements in this direction in public schools and universities remained haphazard, though Cambridge overtook Oxford in institutionalising scientific teaching.[8]

---

[5] Levine,[261] p. 111; Manuel,[139] ch. 13; Lyons,[95] chs. 4–7.
[6] Flamsteed,[162] p. xxxvi; Bernard Smith, *European vision and the South Pacific 1768–1850* (London, 1960), p. 15 and ch. 1 passim. See also above, pp. 133–4.
[7] Hahn,[99] chs. 2, 3, 5; F. B. Artz, *The development of technical education in France 1500–1850* (Cambridge, Mass., 1966), ch. 2.
[8] Hans,[320] chs. 2–3; Ashley-Smith,[321] pp. 246–50; for a more sceptical view, Vincent,[214] esp. pp. 218–19; Frank,[132] pp. 240, 263–4; John Gascoigne, 'The emergence of the Cambridge mathematical tripos', paper delivered at a colloquium on the history of academic disciplines, Middlesex Polytechnic, November 1979.

In technology, too, the contribution of scientists continued to be disappointing, as protracted debate about the relations between science and the Industrial Revolution has revealed. Intentions remained as good as ever: in 1754 a new body, the Society of Arts, was set up to promote this part of the Royal Society's original programme, an implicit recognition of the dissociation of science and technology discussed in chapter 4. But, after 1700 as before, hopes and achievement remained far apart: historians of this century as of its predecessor have been over-impressed by descriptive efforts in encyclopaedias and the like, which rarely affected practice, and have been prone to take aspirations too literally instead of laboriously ascertaining what came of them.[9] Science has also often been misunderstood by presuming the dominance of economic motives.

As a cause and a result of this, there was much continuity in the social and occupational basis of natural philosophy, and the virtuoso element remained strong. Science was still predominantly associated with the leisured and professional establishment, with university-educated men who gathered in metropolitan circles or their provincial reflections, sharing common assumptions and ideals.[10] Some have even discerned a general intellectual torpor in the mid eighteenth century, perhaps resulting from a sort of complacent atavism in a narrow community. Others, however, have been less harsh, seeing this continuity not as an impediment but a useful aid to the growth of knowledge.[11]

But, at the same time, there were important changes in science's audience which were ultimately to alter its role. In the eighteenth century science was popularised on a scale that makes the hesitant beginnings during the Restoration pale into insignificance, by entrepreneurs who achieved much in a field in which they rightly felt themselves to be pioneers. Thus public lectures on scientific subjects – often by itinerant lecturers in the provinces – became common as the appetite for natural philosophy spread through society, part of a wider process by which the appetite for self-improvement and entertainment of a richer and more leisured public was served by a whole

[9] Hudson and Luckhurst,[322] esp. p. 58. See the studies listed on p. 209, below, and also A. E. Musson, ed., *Science, technology and economic growth in the eighteenth century* (London, 1972).

[10] See Roy Porter, 'Gentlemen and geology: the emergence of a scientific career, 1660–1920', *Historical Journal*, 21 (1978), 811–15.

[11] Stuart Piggott, *William Stukeley* (Oxford, 1950), pp. ix–x, 183; Allen,[317] pp. 15–16; Porter,[46] pp. 91–4, 112–18.

range of novel devices. There was also an ever-increasing market for popularising books and periodicals and cheap scientific instruments.[12] This growing availability of scientific ideas gradually affected research, for statistical study has revealed a broadening of the social and educational origins of scientists during the eighteenth century.[13]

Popularisation was not least important in transforming the role of provincial science. The eighteenth century saw the emergence of a real vitality in the provinces which contrasts with the precarious state of affairs surveyed in chapter 3. This has been most fully studied in the Midlands and the North in the period of industrialisation, which has sometimes led to a mistaken belief that it was directly connected with the Industrial Revolution. But it is now clear that science flourished as much in Bath and elsewhere in the South of England as it did in the North: it reflected an attempt to espouse metropolitan, Enlightenment values, a symptom of growing self-consciousness and a commitment to progressive and civilised ideals, rather than an attempt to pursue immediate utility. It did, however, often involve a sense of the supremacy of scientific knowledge and a belief in its potential which had previously been lacking among all but a few.[14]

The ultimate implication of this growth of provincial science was a reshaping of metropolitan science in the years around 1800 by new institutions consciously different from the old-style Royal Society. This was connected partly with the diversification necessitated by the sheer increase of activity in the various branches of natural knowledge, partly with a serious and widespread commitment to the potential of science in all areas of human endeavour. It was in the early nineteenth century that the word 'scientist' was coined, and the role of science as a virtual national ideology has been cogently argued in relation to the foundation and early history of the Royal Institution, perhaps the most significant of these new scientific bodies.[15] Here one encounters a firm belief that scientific methods and standards could achieve anything to which the Western World has been the heir ever since; it brought with it a now familiar cult of

---

[12] See the studies listed on p. 216 below. There is also information in Meyer[150] and Musson and Robinson.[186]

[13] Hans,[320] pp. 32–5.   [14] Porter[327]; Thackray,[324] esp. p. 709.

[15] Sydney Ross, '*Scientist*: the story of a word', *Annals of Science*, 18 (1962), 65–86; Berman,[325] introduction, ch. 4.

expertise and professionalism which affected even government to an extent previously unparalleled.

This was part of the triumph throughout Europe of a new orthodoxy in which science reached genuine cultural dominance: one author, Joseph Ben-David, has evoked a 'scientistic' as against a 'scientific' movement in which scientifically-based norms and methods affected every field of knowledge, from economics to political thought.[16] An index of the percolation of science into eighteenth-century thought is provided by the literature of the era, where a virtual obsession with the findings of the new science was accompanied by a seepage of scientific ideas into writers' vocabulary and imagery: a 'triumph of science' has also been discerned in European art.[17] At the same time, attitudes towards scenery that had originated in scientific circles in the late seventeenth century became commonplace, and the predominance of the natural theology championed by seventeenth-century scientists is shown by its diffusion even into parochial sermons.[18]

Such instances could be multiplied, and their breathless juxtaposition obscures the varying speed and efficacy with which science penetrated different fields. What is important about them, however, is their negative implication for the seventeenth century, when related developments were in their infancy and the role of science easily overestimated. Many authors have anachronistically assumed a cultural hegemony for science too early and even Joseph Ben-David is perhaps premature in acclaiming his 'scientistic' revolution. In the seventeenth century a 'scientific ideology' hardly existed except among unrepresentative enthusiasts, and modern scholars who have implied the contrary have been misled. Institutional and governmental support for natural philosophy was haphazard, enthusiasm sporadic, hostility widespread, and apathy and indifference dominant. Scientific enterprise in the age of the early Royal Society may have had an astounding vitality, but it is only in this setting that the hopes, achievements, failures and anxieties of the new philosophers can be properly understood.

[16] Ben-David,[335] ch. 5; Letwin,[216] ch. 8; S. R. Letwin, *The pursuit of certainty* (Cambridge, 1965).

[17] In addition to Nicolson[328] and Jones,[329] see Donald Davie, *The language of science and the language of literature 1700–40* (London, 1963); Smith, *European vision and the South Pacific*, p. 254.

[18] Nicolson,[330] chs. 7–8; Sykes,[12] pp. 176f.

Appendix

mmmmmmmmmmmmmmmmmmmmmmmmmmmmmmmmmmmmmmmmmmmmmmmmmmmmmmmmmmmmmmmmmmmmmmmmmm

The pictorial frontispiece to Thomas Sprat's *History of the Royal Society* is well-known and has frequently been reproduced. It is often and rightly taken as a visual expression of the leading values of Restoration science, its Baconian inspiration and its commitment to the contemporary establishment. But, though it forms a highly appropriate introduction to Sprat's text, the print's relationship to his volume is puzzling. It is not found in all copies of the book and there is no reason to think that it has been removed from those from which it is absent: it has sometimes been claimed that it was included only in large-paper copies, but it is also present in some of ordinary size.[1] Moreover in these the plate is far too large for the format of the book, so that it has had to be cropped and folded to fit. It is also oddly placed, since the book has another frontispiece, an engraving of the Society's arms: this faces the title-page, and in most copies the pictoral print is placed rather uncomfortably between them. Hitherto no reason for these anomalies has been suggested, but their explanation is clear. The plate was not prepared for Sprat's book but for another volume with overlapping intentions which was never published. This work was devised by John Beale.

Beale has already appeared in this book as a garrulous country correspondent of Oldenburg; he also wrote frequently to Boyle and Evelyn. His location in rural Somerset and his diffuse and prolix epistolary style have led some modern scholars to dismiss him as an eccentric outsider to scientific circles.[2] But – for all his oddity – he was taken seriously at the time, and his sedulity as a writer of letters and papers assured him a surprisingly major role in the early Royal Society, both in suggesting projects and stimulating co-operation. It was he who suggested the formation of the Society's 'Georgical Committee' in 1664, thus provoking the Society to establish the committees on all aspects of its interests that have already been referred to.[3] Through him such important contributions from provincial naturalists as the observations on the tides at Bristol used by Newton reached the Society; and to Beale, who frequently contributed to the *Philosophical Transactions*,

[1] Geoffrey Keynes, *John Evelyn: a study in bibliophily with a bibliography of his writings* (second edition: Oxford, 1968), p. 284.   [2] Oldenburg,[121] I, 320n.
[3] Birch,[118] I, 402–3; see above, pp. 37, 92–3, 94.

Oldenburg sent for comment and improvement his 'Dedication' to volume IV of that journal in 1670.[4]

It is hence not surprising that Beale was taken seriously when he put forward a project for disseminating esteem for the Royal Society. He discussed the matter with Sir Robert Moray and other prominent Fellows in 1664, and he subsequently claimed to have been encouraged in it by Oldenburg, Hooke and Boyle.[5] His plan partook of his general conviction that the spread of interest in the new science and related topics depended on vigorous propaganda, on what he called 'the right management of fame', citing Bacon's unfinished 'Essay on Fame' (first published in 1657) and the experience of his former mentor, Sir Henry Wotton, one-time Provost of Eton, on the ways in which the spread of information could be manipulated. Indeed apart from direct references to 'the conduct of fame', the need for the active dissemination of ideas and improvements through those with influence in the country at large forms a constant theme in Beale's letters, as with his hope to see cider-making encouraged all over England, which he put to the Royal Society in 1663.[6]

His scheme for promoting the Royal Society involved the preparation of what he referred to as 'Lord Bacons Elogyes &c', believing that 'our way to support our owne Enterprise is to devise all wayes to revive Lord Bacons lustre'. It was apparently originally intended as an engraved broadsheet headed by a picture of Bacon, but Beale seems subsequently to have thought more in terms of a printed book which would have included 'A briefe Representation in sumary heades pointing out The Progresse of Solid Learning Since my Lord Bacon invited the Ingenuous To addict to faythfull experiments And more especially since it was countenanced And undertaken by the Royall Society'.[7] This too, however, was to have had an engraved illustration, and it was in connection with this that Beale tried to interest John Evelyn in his plan. Evelyn evidently agreed to prepare a design, and Beale's letters to him provide such details as are known of the project and its ultimate abandonment. Though some points remain frustratingly vague due to lacunae in the correspondence, enough evidence is available to reconstruct the outlines of this proposed defence of the new science, discrete from but similar to those actually brought to fruition by Sprat and Glanvill.

The idea is first mentioned in November 1664; Beale elaborated his plans in more detail in 1665; a passing reference in a letter of 1666 shows that the

---

[4] Beale to Oldenburg, 23 May 1668, Oldenburg,[121] IV, 424–6; see above, p. 65; Beale to Oldenburg, ?Feb. 1670, Oldenburg,[121] VI, 474, and ibid, xxv.

[5] Beale to Evelyn, Nov. 1664, 29 April 1667, Evelyn, *Corr.*, nos. 46, 57; Oldenburg to Boyle, 17 Nov. 1664, Oldenburg,[121] II, 312.

[6] Beale to Boyle, 21 Nov. 1663, 26 June 1682, Boyle,[71] VI, 374–6, 447; Bacon,[69] VI, 519–20; see also Beale to Evelyn, 13 Dec. 1662, 24 Aug. 1667, Christ Church, Oxford, Evelyn Collection, MS 3.1, fol. 35, Evelyn, *Corr.*, no. 62, Beale to Oldenburg, 15 Jan. 1663, 1 June 1667, Oldenburg,[121] II, 6–7, III, 426; Birch,[118] I, 172.

[7] Beale to Evelyn, Nov. 1664, 22 April, 26 April 1665, n.d. [1667], Evelyn, *Corr.*, nos. 46, 47, 48, 54.

project was still alive and it re-emerged to prominence in 1667.[8] The relationship between Beale's scheme and Sprat's *History* is problematic. Sprat was commissioned to write his book in 1663 and part of it was printed in 1664 although various complications delayed its completion until 1667.[9] Yet Moray and others apparently encouraged Beale with his enterprise in 1664 despite its overlap with Sprat's work. Subsequently, when Sprat's book failed to appear, Beale pursued his project and perhaps expanded its scale, evidently believing that Sprat had abandoned his task and that his own expedient was essential to assure the Royal Society the publicity that he was convinced it needed.[10] But in April 1667 Beale learnt that Sprat's *History* was on the verge of publication; from what he heard about it he was apprehensive that it would 'expunge the best, the hearte, & allmost the Totall of my designe', which he therefore decided to abandon, instead looking forward enthusiastically to the appearance of the *History*. By this time, however, Evelyn had completed a design for Beale's use in his book, and Beale suggested that it should instead be inserted either into Sprat's volume or into Glanvill's impending defence of the new science.[11] In the event it was used for Sprat's *History*, and it was possibly because this suggestion was only taken up when publication was well-advanced that some copies were issued without it.

When Beale suggested its diversion to Sprat's book the print had evidently been designed but not engraved (it is dated 1667) and it is possible that Evelyn modified it for Sprat's use. But it is more likely – not least in view of Evelyn's habitual economy with his work – that the actual design done for Beale was re-used, as is further suggested by the print's excessive size for Sprat's volume. Though the design is well-suited to Sprat's *History*, it would have fitted the work that Beale proposed equally well. Indeed some elements in the design seem to take up suggestions of Beale, though others probably originated with Evelyn, to whom Beale left the responsibility for its details.[12]

It is clear from Beale's letters to Evelyn that the basic conception of a pictorial celebration of Francis Bacon and the Stuart monarchy in an architectural setting was his. His notion of showing Bacon's apotheosis was not adopted, and neither was the idea (probably proposed by Evelyn, but eagerly elaborated by Beale) of including columns of the major classical orders to symbolise the three Stuart kings: on the print, Charles II's bust appears on a nondescript pedestal. But the final design did take up Beale's wish to show Bacon seated in his robes as Lord Chancellor, thus reflecting his anxiety for the Royal Society to use Bacon 'to Angle for the young Lawyers, who are the generality of the English Gentry' whose apathy he so

[8] Beale to Evelyn, Nov. 1664, 22 April, 26 April 1665, 10 Feb. 1666, n.d. [1667], Evelyn, *Corr.*, nos. 46, 47, 48, 50, 54.

[9] Sprat,[92] pp. xiii–xiv.

[10] Beale to Evelyn, n.d. [1667], Evelyn, *Corr.*, no. 54.

[11] Beale to Evelyn, 29 April 1667, Evelyn, *Corr.*, no. 57–8; Beale to Oldenburg, 1 June 1667, Oldenburg,[121] III, 426–7.

[12] Beale to Evelyn, 22 April 1665, Evelyn, *Corr.*, no. 47.

resented.[13] Moreover there can be no doubt that the prominent winged figure of Fame with her characteristic trumpet, who crowns Charles II with a laurel wreath, is included to embody Beale's preoccupation with 'the Mastery of Fame'.[14]

Other details, however, may well be Evelyn's own contributions to the design, including the bust of Charles II and the sensitive portrait of the President, William, Viscount Brouncker. The figures sit under an architectural composition which (it has been suggested) might be meant to recall the tiled piazzas on the north and south sides of the Green Court at Gresham College, though it is likelier to be purely imaginary.[15] Above them are the arms of the Society, while to their left is a table bearing the mace presented by the King and the Society's charter and statutes. Beyond that is a bookcase, some of the volumes in which are identified as works by Bacon, by heroes of the new science like Copernicus and Harvey, and by Fellows of the Royal Society. On both sides of the figures are placed or hung a selection of scientific instruments, many of them up-to-date ones with which Fellows of the Society were currently experimenting. In the background an idealised landscape holds a telescope, perhaps intended as the 35-foot instrument presented by Sir Paul Neile to Gresham College in 1658, while beyond is a building that could be meant to suggest 'Solomon's House' in Bacon's *New Atlantis*.[16] The plate was engraved by Wenceslaus Hollar.

[13] Beale to Evelyn, Nov. 1664, 22 April 1665, January 1671, Evelyn, *Corr.*, nos. 46, 47, 108.
[14] Beale to Oldenburg, 1 June 1667, Oldenburg,[121] III, 426. There is no need to see a specific Rosicrucian allusion in this figure, as suggested in F. A. Yates, *The Rosicrucian enlightenment* (London, 1972), pp. 191–2.
[15] Keynes, *John Evelyn*, p. 184.
[16] Hartley,[97] p. 159 and plate 19.

Bibliographical essay

mmmmmmmmmmmmmmmmmmmmmmmmmmmmmmmmmmmmmmmmmmmmmmmmmmmmmmmmmmmmmmmmmmmmmmmmm

This essay supplements the book by providing a critical introduction to existing scholarship on Restoration science and its context. Its arrangement follows the chapters of the book, but its coverage naturally differs: it can offer little on many themes developed at length in the text on the basis of original research, whereas some controversies on rather unimportant points are dealt with fully here but make proportionately little appearance in the book itself. Apart from secondary literature, some indication has been given of primary sources which are easily available either in modern editions or reprints. In the case of reprints, place and date of publication has not been given unless important editorial matter is included.[1] For all other modern books, however, full details of imprint have been provided as an aid to locating them. The sequence of bold, raised numerals provides a code for citation in the footnotes. Books and articles cited here may be located by author through the index.

INTRODUCTION

The Restoration has often been seen as a postscript to the revolutionary upheavals of the mid seventeenth century, and has attracted less attention from modern historians than the spectacular developments that preceded it. Of late, however, political historians have taken more interest in the unsettled nature of the era. This has been accompanied by an interest in economic developments that can be seen as paving the way to the Industrial Revolution though social history still, on the whole, lags behind. Two recent books conveniently introduce modern views of the Restoration: Joan Thirsk,[1] ed., *The Restoration* (London: Longman, 1976) is a useful anthology surveying a number of themes, while J. R. Jones,[2] ed., *The restored monarchy 1660–88* (London: Macmillan, 1979) has essays by various scholars which give up-to-date analyses of different topics and also provide helpful guides to reading. On the period after 1688 this volume has a slightly older equivalent in Geoffrey Holmes,[3] ed., *Britain after the Glorious Revolution 1689–1714* (London: Macmillan, 1969), while on the preceding years

---

[1] For a guide to reprints in print see Christa Gnirss, ed., *International bibliography of reprints*, vol. I, Books and serials, pt. 1–3 (Munich: Verlag Dokumentation, 1976), which is supplemented by the quarterly *Bulletin of Reprints*.

there is another volume in the same series, G. E. Aylmer,[4] ed., *The Interregnum: the quest for a settlement 1646–60* (London: Macmillan, 1972).

The two books which together give the best indication of the political character of the Restoration are J. H. Plumb,[5] *The growth of political stability in England 1675–1725* (London: Macmillan, 1967) and J. R. Western,[6] *Monarchy and revolution: the English state in the 1680s* (London: Blandford, 1972). For political narrative, the most up-to-date survey is J. R. Jones,[7] *Country and court: England 1658–1714* (London: Arnold, 1978), while two useful older works are David Ogg,[8, 9] *England in the reign of Charles II*, second edition (Oxford: Clarendon Press, 1956) and *England in the reigns of James II and William III* (Oxford: Clarendon Press, 1955).

On the Restoration church, the best introduction is a volume edited by G. F. Nuttall and Owen Chadwick,[10] *From uniformity to unity 1662–1962* (London: S.P.C.K., 1962), especially the essays by Anne Whiteman and Roger Thomas; on politico-religious attitudes, John Miller,[11] *Popery and politics in England 1660–88* (Cambridge: Cambridge University Press, 1973) has some interesting chapters. The later years of the century are the subject of an important essay by G. V. Bennett in Holmes,[3] while a longer perspective is provided by Norman Sykes,[12] *From Sheldon to Secker: aspects of English church history 1660–1768* (Cambridge: Cambridge University Press, 1959).

For economic and social history, by far the best introduction is Charles Wilson,[13] *England's apprenticeship 1603–1763* (London: Longman, 1965). The expansion of trade is surveyed in Ralph Davis,[14] 'English foreign trade, 1660–1700', *Economic History Review*, second series 7 (1954), 150–66, while on the 'growth of leisure', mainly after 1700, though with some reference to the late seventeenth century, see J. H. Plumb,[15] *The commercialisation of leisure in eighteenth-century England* (University of Reading, 1973). On the neglected field of social history there are some remarks in Lawrence Stone,[16] 'Social mobility in England, 1500–1700', *Past and Present*, 33 (1966), 16–55, and an article by W. A. Speck in Holmes.[3] Much information on economic and social themes is available from Joan Thirsk and J. P. Cooper,[17] eds., *Seventeenth-century economic documents* (Oxford: Clarendon Press, 1972), while for a more general selection of sources there is Andrew Browning,[18] ed., *English Historical Documents*, 8 (1660–1714) (London: Eyre and Spottiswoode, 1953). The most useful bibliography is W. L. Sachse,[19] *Restoration England 1660–89* (Cambridge: Cambridge University Press, 1971).

1: RESTORATION SCIENCE: ITS CHARACTER AND ORIGINS
Francis Bacon and the Scientific Revolution

For general surveys of Restoration science in its European context, see A. R. Hall,[20] *From Galileo to Newton 1630–1720* (London: Collins, 1963) and R. S. Westfall,[21] *The construction of modern science: mechanisms and mechanics* (New York: Wiley, 1971; Cambridge: Cambridge University Press, 1977). A recent introduction to science before *c.*1650 which lays more stress on mystical ideas is A. G. Debus,[22] *Man and nature in the renaissance* (Cambridge: Cambridge University Press, 1978), while the background up to and including Bacon is briefly surveyed in W. P. D. Wightman,[23] *Science in a renaissance society* (London: Hutchinson, 1972).

The significance of Baconianism is perhaps most uncritically asserted in Margery Purver,[24] *The Royal Society: concept and creation* (London: Routledge, 1967), whose arguments have not been widely accepted. At the opposite extreme, an account which downplays the role of experiment in conceptual change is Alexandre Koyré,[25] *Galileo studies* (English translation: Hassocks: Harvester, 1978). For more of a compromise position see T. S. Kuhn,[26] 'Mathematical versus experimental traditions in the development of physical science' in *The essential tension* (Chicago: Chicago University Press, 1978), pp. 31–65 (previously printed in *Annales*, 30 (1975) and *Journal of Interdisciplinary History*, 7 (1976–7)). Moderate assessments of Baconianism will be found (among other things) in B. C. Teague,[27] 'The origins of Robert Boyle's philosophy' (Cambridge Ph.D. thesis, 1971) and Paul Wood,[28] 'Francis Bacon and the "experimentall philosophy": a study in seventeenth-century methodology' (London M.Phil. thesis, 1978); this may be supplemented on one of the figures it deals with by the interesting case-study of Charles Webster,[29] 'Henry Power's experimental philosophy', *Ambix*, 14 (1967), 150–78.

Important recent studies of scientific epistemology and method include R. S. Westfall,[30] 'Unpublished Boyle papers relating to scientific method', *Annals of Science*, 12 (1956), 63–73, 103–17; Maurice Mandelbaum,[31] *Philosophy, science and sense perception* (Baltimore: Johns Hopkins Press, 1964); A. R. and M. B. Hall,[32] 'Philosophy and natural philosophy: Boyle and Spinoza' in *Mélanges Alexandre Koyré* (Paris: Hermann, 1964), II, 241–56; M. B. Hesse,[33] 'Hooke's philosophical algebra', *Isis*, 57 (1966), 67–83; Laurens Laudan,[34] 'The clock metaphor and probabilism: the impact of Descartes on English methodological thought, 1650–65', *Annals of Science*, 22 (1966), 73–104; G. A. J. Rogers,[35] 'Descartes and the method of English science', ibid. 29 (1972), 237–55; D. J. Oldroyd,[36] 'Robert Hooke's methodology of science as exemplified by his "Discourse of earthquakes" ', *British Journal for the History of Science*, 6 (1972), 109–30; J. F. McDonald,[37] 'Properties and causes: an approach to the problem of hypothesis in the scientific methodology of Sir Isaac Newton', *Annals of Science*, 28 (1972), 217–33; and J. W. Yolton,[38] *Locke and the compass of human understanding* (Cambridge: Cambridge University Press, 1970). For an interesting side-light on the contemporary faith in mathematical reasoning see L. I. Bredvold,[39] 'The invention of ethical calculus' in *The seventeenth century: studies in the history of English thought and literature from Bacon to Pope by R. F. Jones and others writing in his honor* (Stanford: Stanford University Press, 1951), pp. 165–80.

Of the various 'research traditions' referred to, the following modern accounts may be singled out. By far the best work on medical research in the later seventeenth century is R. G. Frank, jr,[40] *Harvey and the Oxford physiologists: a study of scientific ideas and social interaction* (Berkeley and Los Angeles: University of California Press, 1980); there is related information in L. S. King,[41] *The road to medical enlightenment 1650–95* (London: Macdonald, 1970) and A.G. Debus,[42] ed., *Medicine in seventeenth-century England* (Berkeley and Los Angeles: University of California Press, 1974). On the Paracelsian tradition there is A. G. Debus,[43] *The chemical philosophy* (New York: Science History Publications, 1977), while on seventeenth-century chemistry and Boyle's role in it see T. S. Kuhn,[44] 'Robert Boyle and structural chemistry in the seventeenth century', *Isis*, 43 (1952), 12–36, in addition to Marie Boas [Hall],[45] *Robert Boyle and seventeenth-century chemistry* (Cambridge: Cambridge University Press, 1958). The earth sciences are covered by Roy Porter,[46] *The making of geology* (Cambridge: Cambridge University Press, 1977).

Perhaps the best single work on the Galilean tradition is R. S. Westfall,[47] *Force in Newton's physics* (London: Macdonald, 1971), which may be supplemented by J. A. Bennett,[48] 'Hooke and Wren and the system of the world: some points towards an historical account', *British Journal for the History of Science*, 8 (1975), 32–61; see also F. F. Centore,[49] *Robert Hooke's contributions to mechanics* (The Hague: Nijhoff, 1970). A succinct introduction to the 'mechanical philosophy' is available in R. H. Kargon,[50] *Atomism in England from Hariot to Newton* (Oxford: Clarendon Press, 1966); there is also Marie Boas [Hall],[51] 'The establishment of the mechanical philosophy', *Osiris*, 10 (1952), 412–541, while a wide-ranging and important cognate study is E. J. Dijksterhuis,[52] *The mechanisation of the world picture* (Oxford: Clarendon Press, 1961). Two other related recent books are A. I. Sabra,[53] *Theories of light from Descartes to Newton* (London: Oldbourne, 1967) and J. L. Heilbron,[54] *Electricity in the seventeenth and eighteenth centuries* (Berkeley and Los Angeles: University of California Press, 1979).

On the vexed question of the 'occult' element in the science of the time there is a full discussion, especially of minor figures, in K. T. Hoppen,[55] 'The nature of the early Royal Society', *British Journal for the History of Science*, 9 (1976), 1–24, 243–73. Earlier developments are best summarised in F. A. Yates,[56] 'The Hermetic tradition in renaissance science' in C. S. Singleton, ed., *Art, science and history in the Renaissance* (Baltimore: Johns Hopkins Press 1967), pp. 255–74, while Keith Thomas,[57] *Religion and the decline of magic* (London: Weidenfeld, 1971) provides an essential background, indicating the commonness of occult pursuits in seventeenth-century English society as a whole.

Newton's alchemical studies and their significance are best approached through R. S. Westfall,[58, 59] 'Newton and the Hermetic tradition' in A. G. Debus, ed., *Science, medicine and society in the Renaissance* (New York: Science History Publications, 1972), II, 183–98, and 'The role of alchemy in Newton's career' in M. L. Righini Bonelli and W. R. Shea, eds., *Reason, experiment, and mysticism in the Scientific Revolution* (London: Macmillan, 1975), pp. 189–232, and B. J. T. Dobbs,[60] *The foundations of Newton's alchemy* (Cambridge: Cambridge University Press, 1975). A call for greater clarity in approaching such matters will be found in J. E. McGuire,[61] 'Neoplatonism and active principles: Newton and the *Corpus Hermeticum*', in R. S. Westman and J. E. McGuire, *Hermeticism and the Scientific Revolu-*

tion (Los Angeles: William Andrews Clark Memorial Library, 1977), pp. 93–142. Another important article is J. E. McGuire and P. M. Rattansi,[62] 'Newton and the "Pipes of Pan"', *Notes and Records of the Royal Society*, 21 (1966), 108–43.

Many important works by scientists are available in modern editions or reprints. Of Newton's *Principia* there is an English translation edited by Florian Cajori[63] (Berkeley: University of California Press, 1934). Robert Hooke,[64, 65] *Micrographia* (London, 1665) and *Posthumous works*, ed. Richard Waller (London, 1705) have both been reprinted in facsimile, as have Joseph Glanvill,[66] *Plus Ultra* (London, 1668) (ed. J. I. Cope: Gainesville: Scholars' Facsimiles and Reprints, 1958) and Robert Plot,[67, 68] *The Natural History of Oxford-shire*, second edition (Oxford, 1705) and *The Natural History of Stafford-shire* (Oxford, 1686). As for standard editions, the following have been used: Francis Bacon,[69] *Works*, ed. James Spedding, R. L. Ellis and D. N. Heath (London, 1857–74), Thomas Hobbes,[70] *English Works*, ed. Sir William Molesworth (London, 1839–45), and Robert Boyle,[71] *Works*, ed. Thomas Birch, second edition (London, 1772).

The legacy of the Interregnum

Profuse information on numerous aspects of Interregnum science, its organisation and context will be found in Charles Webster,[72] *The Great Instauration: science, medicine and reform 1626–60* (London: Duckworth, 1975), by far the best book on the science of the time despite its occasional flaws. This includes up-to-date surveys of scientific groupings in London, Oxford and Cambridge, but it may be supplemented by the following articles: Charles Webster,[73, 74] 'The College of Physicians: "Solomon's House" in Commonwealth England', *Bulletin of the History of Medicine*, 41 (1967), 393–412, and 'New light on the Invisible College: the social relations of English science in the mid-seventeenth century', *Transactions of the Royal Historical Society*, fifth series 24 (1974), 19–42, and R. G. Frank, jr,[75] 'John Aubrey, F.R.S., John Lydall, and science at Commonwealth Oxford', *Notes and Records of the Royal Society*, 27 (1973), 193–217. Webster[72] is the fundamental work on the Hartlib circle, but for a different view of Hartlib and his significance see A. R. Hall,[76] 'Science, technology and utopia in the seventeenth century' in Peter Mathias, ed., *Science and society 1600–1900* (Cambridge: Cambridge University Press, 1972), pp. 33–53. On the overriding religious interests of the Hartlib circle, very much neglected by Webster, there is J. M. Batten,[77] *John Dury: advocate of Christian reunion* (Chicago: Chicago University Press, 1944).

On religious and political radicalism and the ferment of ideas in the late 1640s the most important book is Christopher Hill,[78] *The world turned upside down* (London: Temple Smith, 1972); useful information is also available in Perez Zagorin,[79] *A history of political thought in the English Revolution* (London: Routledge, 1954). On the reaction to such developments and its implications for scientific thought see the slight but suggestive work of P. M. Rattansi,[80, 81] 'Paracelsus and the Puritan Revolution', *Ambix*, 11 (1963), 24–32, and 'The social interpretation of science in the seventeenth century' in Peter Mathias, ed., *Science and society 1600–1900* (Cambridge: Cambridge University Press, 1972), pp. 1–32. Similar views have been developed more fully by J. R. Jacob[82, 83] in 'The ideological origins of Robert Boyle's natural philosophy', *Journal of European Studies*, 2 (1972), 1–21, and *Robert Boyle and the English Revolution* (New York: Burt Franklin, 1977). Jacob not only indicates how Boyle's ideas were shaped by reaction to the sects but also how the Civil War caused Boyle to reject the aristocratic ethos with which he had been brought up in favour of piety, good works and scientific endeavour.

The connection between science and religious moderation is argued in B. J. Shapiro,[84, 85] 'Latitudinarianism and science in seventeenth-century England', *Past and Present*, 40 (1968), 16–41, and *John Wilkins 1614–72: an intellectual biography* (Berkeley and Los Angeles: University of California Press, 1969). This book also makes a case for linking Wilkins's theological standpoint and his epistemological views that echoes Henry van Leeuwen,[86] *The problem of certainty in English thought 1630–90* (The Hague: Nijhoff, 1963), while on the theology of the Latitudinarians, the Cambridge Platonists and other religious parties by far the best account is H. R. McAdoo,[87] *The spirit of Anglicanism: a survey of Anglican theological method in the seventeenth century* (London: Black, 1965). In contrast to the altruism in the affinity of science and moderation stressed by Shapiro, a case for a more aggressive ideology is made in Jacob[83] and J. R. Jacob,[88] 'Restoration, reformation and the origins of the Royal Society', *History of Science*, 13 (1975), 155–76.

Readily available sources include a valuable selection of documents relating to the Hartlib circle in Charles Webster,[89] ed., *Samuel Hartlib and the advancement of learning* (Cambridge: Cambridge University Press, 1970). A. G. Debus,[90] ed., *Science and education in the seventeenth century: the Webster–Ward debate* (London: Macdonald, 1970) contains texts that are revealing about both reformist attitudes and the views of Oxford scientists, while the ideology of the Latitudinarians is expounded in [Simon Patrick],[91] *A Brief Account of the new Sect of Latitude-Men* (London, 1662), which is available in a modern reprint. Thomas Sprat,[92] *The History of the Royal Society of London* (London, 1667) should be consulted in the modern edition of J. I. Cope and H. W. Jones, which includes useful critical apparatus (London: Routledge, 1959). This book has long fascinated historians, who have often taken it too literally. On the work and its author, who was evidently specially enrolled to the Royal Society to produce an eloquent justification of the new science likely to appeal to a wide audience, see Paul Wood,[93] 'Methodology and apologetics: Thomas Sprat's *History of the Royal Society*', *British Journal of the History of Science*, 13 (1980), 1–26.

2: THE SIGNIFICANCE OF THE ROYAL SOCIETY

Various general histories of the Royal Society deal with its early years. The fullest remains C. R. Weld,[94] *A history of the Royal Society* (London: Parker, 1848) while, more recently, Sir Henry Lyons,[95] *The Royal Society 1660–1940* (Cambridge: Cambridge University Press, 1944) is excellent. There is also Dorothy Stimson,[96] *Scientists and amateurs: a history of the Royal Society* (New York: Henry Schuman, 1948) and Harold Hartley,[97] ed., *The Royal Society: its origins and founders* (London: The Royal Society, 1960), which combines an account of the Society's origins and establishment with biographies of various early members.

An account of the Society's early role which also provides a comparative view of European scientific institutions is Martha Ornstein [Bronfenbrenner],[98] *The role of scientific societies in the seventeenth century*, third edition (Chicago: Chicago University Press, 1938). On these foreign institutions the most important books are Roger Hahn,[99] *The anatomy of a scientific institution: the Paris Academy of Sciences 1666–1803* (Berkeley and Los Angeles: University of California Press, 1971) and W. E. K. Middleton,[100] *The experimenters: a study of the Accademia del Cimento* (Baltimore: Johns Hopkins Press, 1971), while Harcourt Brown,[101] *Scientific organisations in seventeenth-century France* (Baltimore: Williams and Wilkins Company, 1934) provides an essential background.

The only recent book-length treatment of the Royal Society is that of Purver,[24] but this is unreliable not only in its simplistic emphasis on the importance of Bacon but also in its one-sided view of the Society's origins. The fullest critique is Charles Webster,[102] 'The origins of the Royal Society', *History of Science*, 6 (1967), 106–28, but reference may also be made to a group of articles published in the book's aftermath: P. M. Rattansi,[103] 'The intellectual origins of the Royal Society', Christopher Hill,[104] 'The intellectual origins of the Royal Society – London or Oxford?' and A. R. and M. B. Hall,[105] 'The intellectual origins of the Royal Society – London and Oxford', *Notes and Records of the Royal Society*, 23 (1968), 129–43, 144–56 and 157–68. It is now clear that it is artificial to seek specifically 'Oxford' or 'London' roots for a Society that owed something to both.

For more recent detailed work on the Society see R. G. Frank, jr[106] 'Institutional structure and scientific activity in the early Royal Society' in *Proceedings of the fourteenth congress of the history of science (1974)* (Tokyo, 1975), IV, 82–101, a computerised analysis of the Society's activity and its changes. On its institutional life and weaknesses there is Michael Hunter,[107] 'The social basis and changing fortunes of an early scientific institution: an analysis of the membership of the Royal Society, 1660–85', *Notes and Records of the Royal Society*, 31 (1976), 9–114;[2] see also R. K. Bluhm,[108] 'Remarks on the Royal Society's finances, 1660–1768', ibid. 13 (1958), 82–103. For a view of the early Royal Society virtually as a social club, see Quentin Skinner,[109] 'Thomas Hobbes and the nature of the early Royal Society', *Historical Journal*, 12 (1969), 217–39. On the Society's relations with Gresham College see Ian Adamson,[110] 'The Royal Society and Gresham College, 1660–1711', *Notes and Records of the Royal Society*, 33 (1978), 1–21. The background to this is provided by his[111] 'The foundation and early history of Gresham College, London, 1596–1704' (Cambridge Ph.D. thesis, 1975), which gives a fuller view of the college's history than the celebrated

---

[2] References to this work in the form 'F127' denote entries in the catalogue of Fellows, pp. 78–114.

article by F. R. Johnson,[112] 'Gresham College: precursor of the Royal Society', *Journal of the History of Ideas*, 1 (1940), 413–38.

On Oldenburg and his correspondence see M. B. Hall,[113, 114] 'Oldenburg and the art of scientific communication', *British Journal for the History of Science*, 2 (1965), 277–90, and 'The Royal Society's role in the diffusion of information in the seventeenth century', *Notes and Records of the Royal Society*, 29 (1974), 173–92. The transatlantic connections of Oldenburg and his successors are illuminated by R. P. Stearns,[115] *Science in the British colonies of America* (Urbana: University of Illinois Press, 1970). On the *Philosophical Transactions* (1665 et seq.) (of which, incidentally, there is a modern reprint, as there is of the *Philosophical Collections* which Hooke brought out from 1679 to 1682 while the *Transactions* were suspended) see E. N. da C. Andrade,[116] 'The birth and early days of the *Philosophical Transactions*', *Notes and Records of the Royal Society*, 20 (1965), 9–27. For institutions formed under the Society's influence there is an excellent account of the Dublin Philosophical Society by K. T. Hoppen,[117] *The common scientist in the seventeenth century* (London: Routledge, 1970), which also provides information about the Oxford Philosophical Society and about attempts to form similar groups elsewhere.

Among sources, Sprat[92] is essential; also very valuable is Thomas Birch,[118] *The history of the Royal Society of London* (London, 1756–7), which, despite its title, merely comprises a transcript of the Society's minutes up to 1687 in all their – often trivialising – detail, with obituaries of prominent Fellows inserted. The Johnson Reprint Corporation edition (New York, 1968) contains a useful introduction and a partial index by A. R. Hall, while a full index of names is now available in G. E. Scala,[119] 'An index of proper names in Thomas Birch, *The history of the Royal Society* (London, 1756–7)', *Notes and Records of the Royal Society*, 28 (1974), 263–329. The minutes of the Oxford Philosophical Society, 1683–90, are printed in R. T. Gunther,[120] ed., *Early science in Oxford*, vol. IV ('The Philosophical Society') (Oxford: for the subscribers, 1925).

The correspondence of Henry Oldenburg[121] is at present being published in full by A. R. and M. B. Hall (Madison and Milwaukee: University of Wisconsin Press, 1965–73, and London: Mansell, 1975–   ). Other important published correspondences include Abraham Hill,[122] *Familiar letters* (London: for W. Johnston, 1767), S. J. Rigaud,[123] ed., *Correspondence of scientific men of the seventeenth century* (Oxford: Oxford University Press, 1841), John Ray,[124, 125] *Correspondence*, ed. E. Lankester (London: Ray Society, 1848) and *Further correspondence*, ed. R. W. T. Gunther (London: Ray Society, 1928), John Evelyn,[126] *Diary and correspondence*, ed. William Bray [and John Forster], vol. III (London: Colburn, 1852) and Isaac Newton,[127] *Correspondence*, ed. H. W. Turnbull, J. F. Scott, A. R. Hall and Laura Tilling (Cambridge: Cambridge University Press, 1959–77). Many of Boyle's letters are printed in Boyle,[71] but a guide to those that are not is available in R. E. W. Maddison,[128] 'A tentative index of the correspondence of the honourable Robert Boyle, F.R.S.', *Notes and Records of the Royal Society*, 13 (1958), 128–201. For the correspondence of Oxford scientists see R. T. Gunther,[129, 130] ed., *Early science in Oxford*, vols. XII ('Dr Plot and the correspondence of the Philosophical Society of Oxford') and xiv ('Life and letters of Edward Lhwyd') (Oxford: for the subscribers, 1939, 1945).

3: THE SCIENTIFIC COMMUNITY

On education, an up-to-date survey of the background is David Cressy,[131] 'Educational opportunity in Tudor and Stuart England', *History of education quarterly*, 16 (1976), 301–20, while there is much information in R. G. Frank, jr,[132] 'Science, medicine and the universities of early modern England: background and sources', *History of science*, 11 (1973), 194–216, 239–69. Frank includes comments on the shared attitudes of researchers like those in the text, but on the whole the definition of the scientific community has gone largely by default, by a presumption that its boundaries are in some way obvious. One study of a related question – with which I do not wholly agree – is J. L. Axtell,[133] 'Locke, Newton and the two cultures' in J. W. Yolton, ed., *John Locke: problems and perspectives* (Cambridge: Cambridge University Press, 1969), 165–82.

Otherwise the materials for generalisation on such matters have to be sought in the biographies of individual scientists. Authoritative and up-to-date details about all major scientists and some minor ones will be found in C. C. Gillespie,[134] ed. in chief, *Dictionary of scientific biography* (New York: Scribner's, 1970–80), while the following books and theses may be singled out: C. E. Raven,[135] *John Ray, naturalist*, second edition (Cambridge: Cambridge University Press, 1950), Margaret 'Espinasse,[136] *Robert Hooke* (London: Heinemann, 1956), J. I. Cope,[137] *Joseph Glanvill, Anglican apologist* (St Louis: Washington University Studies, 1956), Maurice Cranston,[138] *John Locke: a biography* (London: Longman, 1957), F. E. Manuel[139] *A portrait of Isaac Newton* (Cambridge, Mass.: Belknap Press, 1968) R. E. W. Maddison,[140] *The life of the honourable Robert Boyle, F.R.S.* (London: Taylor and Francis, 1969), J. A. Bennett,[141] 'Studies in the life and work of Sir Christopher Wren' (Cambridge Ph.D. thesis, 1974), Jeffrey Carr,[142] 'The biological work of Martin Lister' (Leeds Ph.D. thesis, 1974) (broader than its title implies), Michael Hunter,[143] *John Aubrey and the realm of learning* (London: Duckworth, 1975) and L. G. Sharp,[144] 'Sir William Petty and some aspects of seventeenth-century natural philosophy' (Oxford D.Phil. thesis, 1977). The best study of a minor Fellow of the Royal Society is perhaps D. C. Coleman,[145] *Sir John Banks: baronet and businessman* (Oxford: Clarendon Press, 1963).

The social composition of the early Royal Society is analysed in Hunter.[107] As for the virtuoso background, the theory of the movement is dealt with in W. E. Houghton,[146] 'The English virtuoso in the seventeenth century', *Journal of the History of Ideas*, 3 (1942), 51–73, 190–219, and R. L-W. Caudill,[147] 'Some literary evidence of the development of English virtuoso interests in the seventeenth century, with particular reference to the literature of travel' (Oxford D.Phil. thesis, 1976). Various themes relating to the virtuoso audience for science are covered in M. H. Nicolson,[148] *Pepys' 'Diary' and the new science* (Charlottesville: University Press of Virginia, 1965); there is also Aytoun Ellis,[149] *The penny universities: a history of the coffee-houses* (London: Secker and Warburg, 1956) and G. D. Meyer,[150] *The scientific lady in England 1650–1760* (Berkeley and Los Angeles: University of California Press, 1955).

Moving to the scientific concerns of members of lower social classes, there is a case-study of a scientific apothecary in R. P. Stearns,[151] 'James Petiver, promoter of natural science, c.1663–1718', *Proceedings of the American Antiquarian Society*, new series 62 (1952), 243–365. On almanacs Bernard Capp,[152] *Astrology and the popular press: English almanacs 1500–1800* (London: Faber, 1979) extends the pioneering work of M. H. Nicolson,[153] 'English almanacs and the "new astronomy" ', *Annals of Science*, 4 (1939), 1–33. Scientific instruments and their makers are considered in Maurice Daumas,[154] *Scientific instruments of the seventeenth and eighteenth centuries and their makers* (English translation: London: Batsford, 1972) and Nicholas Goodison,[155] *English barometers 1680–1860*, second edition (Woodbridge: Antique Collectors' Club, 1977). For a mine of information on those with mathematical interests in these and other classes see E. G. R. Taylor,[156] *The mathematical practitioners of Tudor and Stuart England* (Cambridge: Cambridge University Press, 1954), though the data in this book is frustratingly under-referenced and occasionally unreliable.

On science in the provinces, the appearance is eagerly awaited of Anthony Turner,[157] 'Natural philosophy and provincial culture in Britain, 1660–1760: a background to eighteenth-century Bath', a paper read at Bath in 1977. Anthony Turner also has in preparation an edition of the correspondence between William Cole and Sir Robert Southwell, which is to be published by the Bristol Records Society and which will throw much light on intellectual life in Restoration Somerset. On Dublin science there is Hoppen,[117] while the Towneley circle is fully studied in Charles Webster,[158] 'Richard Towneley (1629–1707), the Towneley group and seventeenth-century science', *Transactions of the historic society of Lancashire and Cheshire*, 118 (1966), 51–76. For the background of seventeenth-century local culture, see especially Peter Laslett,[159] 'The gentry of Kent in 1640', *Cambridge Historical Journal*, 9 (1948), 148–64. The best account of the book-trade is Graham Pollard,[160] 'The English market for printed books: the Sandars lectures 1959', *Publishing History*, vol. 4 (1978) 7–48.

It may seem invidious to single out sources in this section, but as far as the biography of scientists is concerned one general and one specific work will be mentioned: John Aubrey,[161] *Brief Lives*, ed. Andrew Clark (Oxford: Clarendon Press,1898) and John Flamsteed,[162] 'History of his own Life and Labors' in Francis Baily, *An account of the revd. John Flamsteed* (London: Admiralty Commissioners, 1835), which has been reprinted. On the social relations of science there are the *Diaries* of Samuel Pepys,[163] ed. R. C. Latham and W. Matthews (London: Bell, 1970–6), John Evelyn,[164] ed. E. S. de Beer (Oxford: Clarendon Press, 1955) and Robert Hooke,[165, 166] ed. H. W. Robinson and W. Adams (London: Taylor and Francis, 1935) (1672–80) and R. T. Gunther in *Early science in Oxford*, vol. x ('The life and work of Robert Hooke', pt iv) (Oxford: for the author, 1935), pp. 69–265 (1688–93).

4: UTILITY AND ITS PROBLEMS

The two classic works asserting the close links between science and technology are Boris Hessen,[167] 'The social and economic roots of Newton's
"Principia" ' in Nicholai Bukharin et al., *Science at the cross roads* (1931),
second edition with useful introduction by P. G. Werskey (London: Cass,
1971), pp. 147–212, and R. K. Merton,[168] *Science, technology and society in
seventeenth-century England* (New York: Fertig, 1970) (originally published in *Osiris*, 4 (1938), 360–632). A related case is argued by Edgar
Zilsel,[169, 170] particularly in 'The sociological roots of science', *American
Journal of Sociology*, 47 (1941–2), 544–62, and 'The genesis of the concept
of scientific progress', *Journal of the History of Ideas*, 6 (1945), 325–49; a
more recent echo of this will be found in Paolo Rossi,[171] *Philosophy, technology and the arts in the early modern era* (English translation: New York:
Harper Torchbooks, 1970).

For statements of an opposing viewpoint see Sir George Clark,[172] *Science
and social welfare in the age of Newton*, second edition (Oxford: Clarendon
Press, 1949; reprinted with additions, 1970) and three works by A. R.
Hall[173, 174, 175]: *Ballistics in the seventeenth century* (Cambridge: Cambridge
University Press, 1952), 'The scholar and the craftsman in the Scientific
Revolution' in Marshall Clagett, ed., *Critical problems in the history of
science* (Madison: University of Wisconsin Press, 1959), pp. 3–23, and
'Merton revisited, or science and society in the seventeenth century', *History
of Science*, 2 (1963), 1–16.

On the Royal Society's interest in technology and its decline see especially
Margaret 'Espinasse,[176] 'The decline and fall of Restoration science', *Past
and Present*, 14 (1958), 71–89. On the 'History of Trades' and its significance for the re-alignment of intellectual life (which he perhaps overrates)
there is W. E. Houghton,[177] 'The History of Trades: its relation to seventeenth-century thought', *Journal of the History of Ideas*, 2 (1941), 33–60:
Houghton, however, has little on the practice as against the theory of the
programme, on which there is some information in Margaret Denny,[178] 'The
early program of the Royal Society and John Evelyn', *Modern Language
Quarterly*, 1 (1940), 481–97.

Proper study of the technological interests of Restoration scientists in their
economic context – comparable to the work of Charles Webster[72] on the
Interregnum – is a real desideratum. Beginnings along these lines are made in
Clark,[172] R. V. Lennard,[179] 'English agriculture under Charles II: the evidence
of the Royal Society's "Enquiries" ', *Economic History Review*, 4 (1932–4),
23–45, W. E. S. Turner,[180] 'A notable British contribution to the literature of
glassmaking', *Glass technology*, 3 (1962), 201–13, and Lindsay Sharp,[181]
'Timber, science and economic reform in the seventeenth century', *Forestry*,
48 (1975), 51–86. On naval technology information is available in
Bennett,[141] Sharp,[144] and Margaret Deacon,[182] *Scientists and the sea 1650–
1900: a study of marine science* (London: Academic Press, 1971), while the
early history of Greenwich observatory is covered in E. G. Forbes,[183] *Greenwich observatory*, vol. 1 ('Origins and early history, 1675–1835') (London:
Taylor and Francis, 1975). For the background see J. R. Tanner,[184] *Samuel*

Pepys and the Royal Navy (Cambridge: Cambridge University Press, 1920) and R. G. Albion,[185] *Forests and sea-power: the timber problem of the Royal Navy 1652–1862* (Cambridge, Mass.: Harvard University Press, 1926).

For conflicting estimates of the link between science and economic change compare A. E. Musson and Eric Robinson,[186] *Science and technology in the Industrial Revolution* (Manchester: Manchester University Press, 1969) with Peter Mathias,[187] 'Who unbound Prometheus? Science and technical change, 1600–1800' in P. Mathias, ed., *Science and society 1600–1900* (Cambridge: Cambridge University Press, 1972), pp. 54–80, or A. R. Hall,[188] 'What did the Industrial Revolution in Britain owe to science?' in N. McKendrick, ed., *Historical perspectives* (London: Europa, 1974), pp. 129–51. More specific studies include J. R. Harris,[189] *Industry and technology in the eighteenth century: Britain and France* (University of Birmingham, 1972), A. and N. L. Clow,[190] 'The timber famine and the development of technology', *Annals of Science*, 12 (1956), 85–102, and G. Hollister-Short,[191] 'Leads and lags in late seventeenth-century English technology', *History of Technology*, 1 (1976), 159–83.

On agriculture the best introduction is Joan Thirsk,[192] 'Seventeenth-century agriculture and social change' in Joan Thirsk, ed., *Land, church and people* (Reading: British Agricultural History Society, 1970), pp. 148–77. Improving literature is surveyed in G. E. Fussell,[193] *The old English farming books from Fitzherbert to Tull 1523–1730* (London: Crosby Lockwood, 1947), while there are general accounts of the spread of novel techniques in Wilson[13] and Eric Kerridge,[194] *The agricultural revolution* (London: Allen and Unwin, 1967), though some of this book's claims are controversial: the volume of the 'Cambridge Agrarian history of England and Wales' for this period is currently in preparation. Two case studies of agricultural innovation are Joan Thirsk,[195] 'New crops and their diffusion: tobacco-growing in seventeenth-century England' in C. W. Chalklin and M. A. Havinden, eds., *Rural change and urban growth 1500–1800* (London: Longman, 1974), pp. 76–103, and Frank Emery,[196] 'The mechanics of innovation: clover cultivation in Wales before 1750', *Journal of Historical Geography*, 2 (1976), 35–48.

Among sources, John Evelyn,[197] *Sylva*, has been reprinted from its first edition (London, 1664). In many ways, however, the third edition (London, 1679) is more interesting due to the additions that Evelyn made both to the text and to the preface 'To the Reader', and this has been used in this book. A number of letters and documents relating to Petty's navigational experiments are printed in the Marquess of Lansdowne,[198] ed., *The double bottom or twin-hulled ship of Sir William Petty* (Oxford: Roxburghe Club, 1931), while there is a modern edition of the second volume of Joseph Moxon,[199] *Mechanick Exercises: or the Doctrine of Handy-works* (London, 1677–84), ed. Herbert Davis and Harry Carter (London: Oxford University Press, 1958). Reprints are available of John Ray,[200] *A Collection of English Words* (London, 1674), John Houghton,[201] *A Collection for Improvement of Husbandry and Trade* (London, 1692–1703) and Daniel Defoe,[202] *An Essay upon Projects* (London, 1697).

5: POLITICS AND REFORM

First it is necessary to survey the literature on science and puritanism. The pioneers were Merton[168] and Dorothy Stimson,[203] 'Puritanism and the new philosophy in seventeenth-century England', *Bulletin of the history of medicine*, 3 (1935), 321–34, while a more recent statement of a similar viewpoint is Christopher Hill,[204] *Intellectual origins of the English Revolution* (Oxford: Clarendon Press, 1965). This book stimulated much debate, not least in the pages of *Past and Present;* the most important contributions to the controversy have been usefully reprinted with other relevant articles from that journal and an 'Introduction' in Charles Webster,[205] ed., *The intellectual revolution of the seventeenth century* (London: Routledge, 1974).[3] Of numerous more or less satisfying attempts to survey and redefine the controversy one might single out T. K. Rabb,[206] 'Puritanism and the rise of experimental science in England', *Cahiers d'histoire mondiale*, 7 (1962), 46–67 and D. S. Kemsley,[207] 'Religious influences in the rise of modern science: a review and criticism, particularly of the "protestant-puritan ethic" theory', *Annals of Science*, 24 (1968), 199–226, while new light is shed on the matter in John Morgan,[208] 'Puritanism and science: a reinterpretation', *Historical Journal*, 22 (1979), 535–60.

A Restoration postscript to this debate is provided by attempts to define the ideological affiliations of the new science through the membership of the Royal Society. Whereas Stimson[203] found a significant 'puritan' strain among the Fellows, she seems to have been mistaken. More respect should be given to the arguments of L. S. Feuer,[209] *The scientific intellectual: the psychological and sociological origins of modern science* (New York: Basic Books, 1963), who discerns the dominance of a 'hedonistic-libertarian ethic', and Lotte Mulligan,[210] 'Civil War politics, religion and the Royal Society', *Past and Present*, 59 (1973), 92–116, who stresses the Society's royalist and Anglican composition, though the difficulties of quantitative analysis of such loyalties make all these conclusions suspect.

Moving on to the theme of the chapter and scientists' reformist ambitions, language-planning is well covered by Vivian Salmon,[211, 212] *The works of Francis Lodwick* (London: Longman, 1972) and 'John Wilkins's *Essay* (1668): critics and continuators', *Historiographia linguistica*, 1 (1974), 147–63, and James Knowlson,[213] *Universal language schemes in England and France 1600–1800* (Toronto: Toronto University Press, 1975). On educational reform there is information in Hunter[143] and Caudill,[147] while W. A. L. Vincent,[214] *The grammar schools: their continuing tradition 1660–1714* (London: Murray, 1969) considers demands and projects for change against a background of the general state of contemporary education.

Useful accounts of political arithmetic will be found in Sharp,[144] Clark,[172] Shichiro Matsukawa,[215] 'Origin and significance of political arithmetic', *Annals of the Hitotsubashi Academy*, 6 no. 1 (1955), 53–79, and William Letwin,[216] *The origins of scientific economics: English economic thought, 1660–1776* (London: Methuen, 1963). On Petty's career in politics see

---

[3] Shapiro,[84] 'Espinasse,[176] and Mulligan[210] are all available in this volume.

E. Strauss,[217] *Sir William Petty: portrait of a genius* (London: Bodley Head, 1954) while on Graunt there is D. V. Glass,[218] 'John Graunt and his *Natural and Political Observations*', *Notes and Records of the Royal Society*, 19 (1964), 63–100. For a rather different reading of Petty and Graunt from that given here, which seems to me to put too much stress on their fear of instability rather than their general distaste for a lack of order, see Peter Buck,[219] 'Seventeenth-century political arithmetic: civil strife and vital statistics', *Isis*, 68 (1977), 67–84. The views of Grew – among others – are surveyed in E. A. J. Johnson,[220] *Predecessors of Adam Smith* (New York: Prentice-Hall, 1937).

On science's political affiliations little work has been done apart from J. R. Jacob,[221] 'Restoration ideologies and the Royal Society', *History of Science*, 18 (1980), 25–38, which focuses on the period of the Exclusion crisis. On politics in the 1680s Western,[6] J. R. Jones,[222] *The Revolution of 1688 in England* (London: Weidenfeld, 1972) and John Miller,[223] *James II: a study in kingship* (Hove: Wayland, 1978) are helpful. For the financial side of affairs C. D. Chandaman,[224] *The English public revenue 1660–88* (Oxford: Clarendon Press, 1975) is important, while on impulses towards efficiency in government circles information will be found in Howard Tomlinson's article in Jones[2] and in G. A. Jacobsen,[225] *William Blathwayt: a late seventeenth-century English administrator* (New Haven: Yale University Press, 1932). On changes in the 1690s the best book is perhaps P. G. M. Dickson,[226] *The financial revolution in England: a study in the development of public credit 1688–1756* (London: Macmillan, 1967). Also interesting from the viewpoint of this chapter are Peter Laslett,[227] 'John Locke, the great recoinage and the origins of the Board of Trade: 1695–8' in J. W. Yolton, ed., *John Locke: problems and perspectives* (Cambridge: Cambridge University Press, 1969), pp. 137–64, and G. N. Clark,[228] *Guide to English commercial statistics 1696–1782* (London: Royal Historical Society, 1938).

On the governmental model provided by contemporary France see J. E. King,[229] *Science and rationalism in the government of Louis XIV 1661–83* (Baltimore: Johns Hopkins Press, 1949), though some feel that this work exaggerates the importance of science as against broader factors. On the very limited English analogues to the service to which science was put by the state in France, Forbes[183] deals with Greenwich observatory, while on the Royal Mathematical School E. H. Pearce,[230] *Annals of Christ's Hospital*, second edition (London: Hugh Rees, 1908) still remains in many ways the best account: Bryan Allen,[231] 'The English mathematical schools 1670–1720' (Reading School of Education Ph.D. thesis, 1970) is quite informative but rather slight.

The most important relevant sources are as follows. John Wilkins,[232] *An Essay Towards a Real Character, And a Philosophical Language* (London, 1668) is available in modern reprint. C. H. Hull[233] has edited William Petty's *Economic writings* (Cambridge: Cambridge University Press, 1899); this edition includes Graunt's *Natural and Political Observations . . . upon the Bills of Mortality* (London, 1662), on the controversy concerning the extent of Petty's contribution to which see Glass.[218] More information about Petty's ideas will be found in writings edited by the Marquis of Lansdowne[234, 235]:

The Petty Papers (London: Constable, 1927) and *The Petty-Southwell correspondence 1676–87* (London: Constable, 1928). Pepys's interest in reformist projects is illustrated by Samuel Pepys,[236, 237] *Naval Minutes*, ed. J. R. Tanner (London: Navy Records Society, 1926) and *Tangier papers*, ed. E. Chappell (London: Navy Records Society, 1935). On early scientific voyages see Edmond Halley,[238] *Correspondence and papers*, ed. E. F. MacPike (Oxford: Clarendon Press, 1932).

6: SCIENCE, LEARNING AND THE UNIVERSITIES

Of the numerous accounts of Stubbe's and Casaubon's assaults on the new science and their background perhaps the best is R. H. Syfret,[239a, 239b] 'Some early reaction to the Royal Society', *Notes and Records of the Royal Society*, 7 (1950), 207–58, and 'Some early critics of the Royal Society', ibid. 8 (1950), 20–64. R. F. Jones,[240] *Ancients and moderns: a study of the rise of the scientific movement in seventeenth-century England*, second edition (St Louis: Washington University Studies, 1961) forces the quarrel to an un-natural extent into the context of 'the background of the "Battle of the Books" ' (to quote the book's sub-title in its original edition). A briefer summary of Jones's views is available in R. F. Jones,[241] 'The background of the attack on science in the age of Pope' in J. L. Clifford and L. A. Landa, eds., *Pope and his contemporaries* (Oxford: Clarendon Press, 1949), pp. 96–113, while for a slighter and more bibliographical account of the dispute there is H. W. Jones,[242] 'Mid-seventeenth-century science: some polemics', *Osiris*, 9 (1950), 254–74. J. R. Jacob has in preparation an analysis of Stubbe's ideas on these and other topics, while Casaubon's important unpublished tract on learning is discussed by its discoverer in M. R. G. Spiller,[243] 'Conservative opinion and the new science 1630–80: with special reference to the life and works of Meric Casaubon' (Oxford B.Litt. thesis, 1968).

A good – if superficial – introduction to the Restoration universities will be found in Hugh Kearney,[244] *Scholars and gentlemen: universities and society in pre-industrial Britain 1500–1700* (London: Faber, 1970). On the Interregnum background R. F. Jones,[245] 'The humanistic defence of learning in the mid-seventeenth century' in J. A. Mazzeo, ed., *Reason and the imagination* (London: Routledge, 1962), pp. 71–92, is interesting, while W. T. Costello,[246] *The scholastic curriculum at early seventeenth-century Cambridge* (Cambridge, Mass.: Harvard University Press, 1958) gives detail about teaching which is relevant to this period. Scientific facilities at the universities are now well studied. Frank[132] largely, but not wholly, supersedes the earlier work of Phyllis Allen,[247, 248] 'Medical education in seventeenth-century England', *Journal of the History of Medicine*, 1 (1946), 115–43, and 'Scientific studies in the English universities of the seventeenth century', *Journal of the History of Ideas*, 10 (1949), 219–53, Arthur Rook,[249] 'Medicine at Cambridge, 1660–1760', *Medical History*, 13 (1969), 107–22, and B. J. Shapiro,[250] 'The universities and science in seventeenth-century England', *Journal of British Studies*, 10 no. 2 (1971), 47–82.

On the College of Physicians and its institutional position see Sir George

Clark,[251] *A history of the Royal College of Physicians of London* (Oxford: Clarendon Press, 1964–6). Also relevant are P. M. Rattansi,[252] 'The Helmontian-Galenist controversy in Restoration England', *Ambix*, 12 (1964), 1–23, and T. M. Brown,[253] 'The College of Physicians and the acceptance of iatromechanism in England, 1665–95', *Bulletin of the History of Medicine*, 44 (1970), 12–30.

The best recent study of the appeal to antiquity and the contemporary tradition of ecclesiastical scholarship is G. V. Bennett,[254] 'Patristic tradition in Anglican thought, 1660–1900' in *Oecumenica, jahrbuch für ökumenische Forschung, 1971/2* (Strasbourg: G. Mohn, 1972), pp. 63–85; information will also be found in Sykes[12] and McAdoo.[87] On scholarship see also Harry Carter,[255] *A history of the Oxford University Press*, vol. 1 (Oxford: Clarendon Press, 1975), Michael Hunter,[256] 'The origins of the Oxford University Press', *The Book Collector*, 24 (1975), 511–34 (which makes explicit the scholarly affiliations of Fell's venture) and Adam Fox,[257] *John Mill and Richard Bentley: a study of the textual criticism of the New Testament 1675–1729* (Oxford: Blackwell, 1954). Historical scholarship is surveyed in D. C. Douglas,[258] *English scholars 1660–1730*, second edition (London: Eyre and Spottiswoode, 1951), which is a classic despite its tendency to anachronistic value judgements; J. G. A. Pocock,[259] *The ancient constitution and the feudal law* (Cambridge: Cambridge University Press, 1957) is also important. Archaeological antiquarianism is surveyed in Hunter[143] and M. C. W. Hunter,[260] 'The Royal Society and the origins of British archaeology', *Antiquity*, 65 (1971), 113–21, 187–92, while there is much information on the world of scholarship from a slightly different angle in J. M. Levine,[261] *Dr Woodward's shield: history, science and satire in Augustan England* (Berkeley and Los Angeles: University of California Press, 1977).

On the 'Ancients and Moderns' dispute of the 1690s there is an early and slight essay by R. F. Jones,[262] 'The background of *The Battle of the Books*' in *The seventeenth century: studies in the history of English thought and literature from Bacon to Pope by R. F. Jones and others writing in his honor* (Stanford: Stanford University Press, 1951), pp. 10–40; it was originally published in 1920. A new view of the controversy is in preparation by J. M. Levine, who has already published an important study[263] of 'Ancients, moderns and history: the continuity of English historical writing in the later seventeenth century' in P. J. Korshin, ed., *Studies in change and revolution* (Menston: Scolar, 1972), pp. 43–75. A recent account of Temple's outlook is R. F. S. Borkat,[264] 'Sir William Temple, the idea of progress and the meaning of learning', *Durham University Journal*, 71 (1978), 1–7.

As for sources, Meric Casaubon,[265] *A Letter . . . to Peter du Moulin* (Cambridge, 1669) has been reprinted, while it is to be hoped that his unpublished tract on learning, transcribed in Spiller,[243] will soon appear in print. A number of extracts from Stubbe's attacks on the Royal Society are printed in the notes to Sprat.[92] On science at Oxford, apart from Gunther,[120, 129] there is also information in John Wallis,[266] 'Letter against Mr Maidwell (1700)', ed. T. W. Jackson, in *Oxford Historical Society* 'Collectanea', 1 (1885), 269–337. The best edition of Sir William Temple,[267] 'An Essay upon the Ancient and Modern Learning' is in *Five miscellaneous essays*, ed. S. H. Monk (Ann Arbor: University of Michigan Press, 1963), pp. 37–71, while there are modern reprints of Obadiah Walker,[268] *Of Education, Especially of Young Gentlemen* (Oxford, 1673) and William Wotton,[269] *Reflections upon Ancient and Modern Learning* (London, 1694).

7: ATHEISM AND ORTHODOXY

For brief surveys of Restoration infidelity against a broader background see D. C. Allen,[270] *Doubt's boundless sea: skepticism and faith in the Renaissance* (Baltimore: Johns Hopkins Press, 1964) and G. E. Aylmer,[271] 'Unbelief in seventeenth-century England' in Donald Pennington and Keith Thomas, eds., *Puritans and revolutionaries* (Oxford: Clarendon Press, 1978), pp. 22–46. A more ambitious but flawed study of the period after 1660 is John Redwood,[272] *Reason, ridicule and religion: the age of enlightenment in England 1660–1750* (London: Thames and Hudson, 1976).

On specific unorthodox thinkers T. F. Mayo,[273] *Epicurus in England 1650–1725* (Dallas: the Southwest Press, 1934) remains useful, while the best account of hostility to Hobbes is S. I. Mintz,[274] *The hunting of Leviathan* (Cambridge: Cambridge University Press, 1962). Also relevant is Quentin Skinner,[275] 'The ideological context of Hobbes's political thought', *Historical Journal*, 9 (1966), 286–317, and, though the voluminous literature on Hobbes's thought cannot be surveyed here, a start may be made with the essays in K. C. Brown,[276] ed., *Hobbes studies* (Oxford: Blackwell, 1965). On English reactions to Spinoza see R. L. Colie,[277, 278] 'Spinoza and the early English Deists', *Journal of the History of Ideas*, 20 (1959), 23–46, and 'Spinoza in England, 1665–1730', *Proceedings of the American Philosophical Society*, 107 (1963), 183–219. Recent studies of the early Deists include J. A. Redwood,[279] 'Charles Blount (1654–93), Deism and English free thought', *Journal of the History of Ideas*, 35 (1974), 490–8, and M. C. Jacob,[280] 'John Toland and the Newtonian ideology', *Journal of the Warburg and Courtauld Institutes*, 32 (1969), 307–31.

For a general account of Restoration religious thought see G. R. Cragg,[281] *From puritanism to the age of reason* (Cambridge: Cambridge University Press, 1950), though in many ways a more suggestive approach will be found in the computerised analysis of best-sellers of C. J. Sommerville,[282] *Popular religion in Restoration England* (Gainesville: University Presses of Florida, 1977). A helpful indication of the commonplaces purveyed by divines on non-religious topics is available in R. B. Schlatter,[283] *The social ideas of religious leaders 1660–88* (London: Oxford University Press, 1940), while on the 'reformation of manners' see D. W. R. Bahlman,[284] *The moral revolution of 1688* (New Haven: Yale University Press, 1957). This should be supplemented by Eamon Duffy,[285] 'Primitive Christianity revived: religious renewal in Augustan England', *Studies in Church History*, 14 (1977), 287–300, which also deals with the ecclesiastical traditions considered in chapter 6.

On the culture of 'wit' and its ambivalent relations with science, a revealing document is printed in Dorothy Stimson,[286] 'Ballad of Gresham Colledge', *Isis*, 18 (1932), 103–17, though it is there misattributed and its true significance missed. The most satisfactory account of literary attacks on science is Syfret,[239b] while detail is added by Claude Lloyd,[287] 'Shadwell and the virtuosi', *Publications of the Modern Language Association of America*, 44 (1929), 472–94. An idea of the kind of milieu from which such satires originated is given by J. H. Wilson,[288] *The court wits of the Restoration* (Princeton: Princeton University Press, 1948).

On the problem of heterodoxy for thinkers who liked to consider themselves orthodox see D. C. Kubrin,[289, 290] 'Newton and the cyclical cosmos: providence and the mechanical philosophy', *Journal of the History of Ideas*, 28 (1967), 325–46, and 'Providence and the mechanical philosophy: the creation and dissolution of the world in Newtonian thought' (Cornell Ph.D. thesis, 1968). On Halley, Simon Schaffer,[291] 'Halley's atheism and the end of the world', *Notes and Records of the Royal Society*, 32 (1977), 17–40, is to be preferred to W. R. Albury,[292] 'Halley's ode on the *Principia* of Newton and the Epicurean revival in England', *Journal of the History of Ideas*, 39 (1978), 24–43, which seems to me to go too far on rather little evidence. Particularly useful on contemporary reactions to Locke is J. W. Yolton,[293] *John Locke and the way of ideas* (Oxford: Clarendon Press, 1956), while a useful background work is H. J. McLachlan,[294] *Socinianism in seventeenth-century England* (London: Oxford University Press, 1951).

For a general account of the religious thought of scientists see R. S. Westfall,[295] *Science and religion in seventeenth-century England* (New Haven: Yale University Press, 1958). The best study of the Cambridge Platonists' reaction to materialism is R. L. Colie,[296] *Light and enlightenment: a study of the Cambridge Platonists and the Dutch Arminians* (Cambridge: Cambridge University Press, 1957). On Henry More's attitudes towards Descartes see Arrigo Pacchi,[297] *Cartesio in Inghilterra: da More a Boyle* (Rome: Editori Laterza, 1973) and Alan Gabbey,[298] 'Philosophia Cartesiana triumphata: Henry More 1646–71' (paper delivered at the Western Ontario winter colloquium on 'Cartesianism, 1650–1750', October 1973: forthcoming, San Diego: Austin Hill). Also important is M. E. Prior,[299] 'Joseph Glanvill, witchcraft and seventeenth-century science', *Modern Philology*, 30 (1932), 167–93.

On Boyle there is the recent work of J. R. Jacob,[300, 301] 'Robert Boyle and subversive religion in the early Restoration', *Albion*, 6 (1974) 275–93 (also printed in Jacob[83]), and 'Boyle's atomism and the Restoration assault on pagan naturalism', *Social Studies of Science*, 8 (1978), 211–33. These stress Boyle's fear of subversion, whereas a different perspective on Boyle's thought is offered by J. E. McGuire,[302] 'Boyle's conception of nature', *Journal of the History of Ideas*, 33 (1972), 523–42. Much of value on Boyle, Newton and others will be found in E. A. Burtt,[303] *The metaphysical foundations of modern physical science* (London: Routledge, 1924), while Newton is the subject of F. E. Manuel,[304] *The religion of Isaac Newton* (Oxford: Clarendon Press, 1974).

On the Boyle lectures and their background see J. J. Dahm,[305] 'Science and apologetics in the early Boyle lectures', *Church History*, 39 (1970), 172–86, and especially M. C. Jacob,[306, 307] 'The church and the formulation of the

Newtonian world-view', *Journal of European Studies*, 1 (1971), 128–48, and *The Newtonians and the English Revolution 1689–1720* (Hassocks: Harvester, 1976). A more recent article which brings together her ideas and her husband's is J. R. Jacob and M. C. Jacob,[308] 'The Anglican origins of modern science: the metaphysical foundations of the Whig constitution', *Isis*, 71 (1980), 251–67: but this goes further than the earlier writings of either author in overemphasising the significance of a crypto-Republican tradition of free thought at a time when heterodoxy was more diffuse. For a corrective to M. C. Jacob's neglect of orthodox hostility to Newtonianism, see Geoffrey Holmes,[309] 'Science, reason and religion in the age of Newton', *British Journal for the History of Science*, 12 (1979), 164–71, C. B. Wilde,[310] 'Hutchinsonianism, natural philosophy and religious controversy in eighteenth-century Britain', *History of Science*, 18 (1980), 1–24, and L. Stewart,[311] 'Samuel Clarke, Newtonianism and the divided universe of post-Revolutionary England', *Journal of the History of Ideas*, 41 (1980).

The following sources may be mentioned. There are modern editions of Thomas Shadwell,[312] *The Virtuoso*, edited by M. H. Nicolson and D. S. Rodes (London: Arnold, 1966), and Thomas Burnet,[313] *The Sacred Theory of the Earth*, introduced by Basil Willey (Fontwell: Centaur, 1965). Ralph Cudworth,[314] *The True Intellectual System of the Universe* has here been used in the edition of J. L. Mosheim, trans. John Harrison (London: for Thomas Tegg, 1845), though a reprint of the first edition (London, 1678) is also available. There are also reprints of Joseph Glanvill,[315] *Sadducismus Triumphatus*, third edition (London, 1689) and Richard Bentley,[316] *Works*, ed. A. Dyce (London, 1836–8).

CONCLUSION

On the popularisation of science, D. E. Allen,[317] *The naturalist in Britain: a social history* (London: Allen Lane, 1976) and J. R. Millburn,[318] *Benjamin Martin: author, instrument-maker and 'country showman'* (Leyden: Noordhoff, 1976) supplement in different ways David Layton,[319] 'The popularisation of science in Great Britain 1650–1800' (London M.Sc. thesis, 1955). For the background to such developments see Plumb.[15] Education is dealt with in Nicholas Hans,[320] *New trends in education in the eighteenth century* (London: Routledge, 1951) and J. W. Ashley Smith,[321] *The birth of modern education: the contribution of the dissenting academies 1660–1800* (London: Independent Press, 1954).

Older work on institutional history includes Derek Hudson and K. W. Luckhurst,[322] *The Royal Society of Arts 1754–1954* (London: Murray, 1954) and R. E. Schofield,[323] *The Lunar society of Birmingham* (Oxford: Clarendon Press, 1963), which overemphasised the directness of links with industrialisation. New directions are indicated by A. W. Thackray,[324] 'Natural knowledge in cultural context: the Manchester model', *American Historical Review*, 79 (1974), 672–709, and Morris Berman,[325] *Social change and scientific organisation: the Royal Institution 1799–1844* (London: Heinemann, 1978). Also important are Roy Porter,[326, 327] 'The Industrial Revolution and the rise of the science of geology' in M. Teich and R. Young, eds., *Changing perspectives in the history of science* (London: Heinemann, 1973), pp. 320–43, and 'Science, provincial culture and public opinion in Enlightenment England', *British Journal for Eighteenth-century Studies*, 3 (1980), 20–46.

For the impact of science on literature see especially M. H. Nicolson,[328] *Newton demands the muse: Newton's 'Opticks' and the eighteenth-century poets* (Princeton: Princeton University Press, 1946) and W. P. Jones,[329] *The rhetoric of science: a study of scientific ideas and imagery in eighteenth-century English poetry* (London: Routledge, 1966). A highly suggestive study is M. H. Nicolson,[330] *Mountain gloom and mountain glory: the development of the aesthetics of the infinite* (Cornell: Cornell University Press, 1963). On changes in religious ideas Leslie Stephen,[331] *History of English thought in the eighteenth century* (1876), ed. Crane Brinton (London: Hart Davis, 1962) is still to be preferred to the bland surveys of R. N. Stromberg,[332] *Religious liberalism in eighteenth-century England* (London: Oxford University Press, 1954) and G. R. Cragg,[333] *Reason and authority in the eighteenth century* (Cambridge: Cambridge University Press, 1964).

Lastly, for the role of science in changing values in the eighteenth century see Paul Hazard,[334] *La crise de la conscience Européenne 1680–1715* (Paris: Boivin, 1935), translated as *The European mind 1680–1715* (London: Hollis and Carter, 1953), and Joseph Ben-David,[335] *The scientist's role in society* (Englewood Cliffs: Prentice-Hall, 1971).

A NOTE ON MANUSCRIPT SOURCES

As is obvious from the notes, manuscript sources have been extensively used in this book. Since the collections – or even sometimes single volumes – involved are often rather disparate in content, it is impossible to give a survey like that attempted for printed materials. It may, however, be useful to indicate what sources are available and where. A partial guide already exists in M. B. Hall,[336] 'Sources for the history of the Royal Society in the seventeenth century', *History of Science*, 5 (1966), 62–76, while more detail on the most important collection will be found in R. K. Bluhm[337] 'A guide to the archives of the Royal Society and to other manuscripts in its possession', *Notes and Records of the Royal Society*, 12 (1956), 21–39.

The extent of the holdings of the Royal Society is itself symptomatic of the Society's role in Restoration science. Much use has been made of the Classified Papers, the main repository of scientific and technical papers submitted to the Society in the late seventeenth and early eighteenth centuries: many of these are of great interest, particularly those of Hooke, Houghton and others, which are referred to in the notes. Such papers were often copied into the Society's Register Book (of which a later copy is available as well as the original), and this occasionally has material that does not survive among the Classified Papers. The Society's minutes up to 1687 are published in Birch,[118] but after 1687 they are only available in the Original Journal Book (or an eighteenth-century copy of this), while minutes of the Society's early committees will be found – together with numerous documents on administrative matters – in the Domestic Manuscripts, especially vol. 5. The account books, after being mislaid for many years, returned to the Society in 1957 and are now accessible there; on these see Hunter[107] and Bluhm[108].

Of the Society's correspondence, the volumes of Early Letters contain the profuse letters to and from Oldenburg that are currently being published in Oldenburg,[121] and various letters addressed to others in the Society's first two decades and the 1690s. Particularly for the decade following Oldenburg's death the originals of letters to and from the Society often do not survive, but there are copies of many of them in the Letter Books (again, both originals and copies survive). For the 1680s and 1690s much correspondence relating to the Society and to science of the time survives among the manuscripts bequeathed to the British Museum by Sir Hans Sloane, whose collection also contained scientific papers of Grew, Hooke, Power, Courten and others. There are also some scientific manuscripts and volumes of Royal Society papers among the Additional Manuscripts of the British Library Reference Division (as the library of the British Museum is now known). For some details of these and of the catalogues that are available of manuscripts at the British Library and the Royal Society, see Hall.[336] Also useful is M. A. E. Nickson,[338] *The British Library: guide to the catalogues and indexes of the Department of Manuscripts* (London: British Library, 1978).

Of the papers of individual scientists, perhaps the most important and least explored are the Boyle Letters and Papers at the Royal Society: the letters are itemised in Maddison[128] and in the general card catalogue at the Royal Society, but a catalogue of the papers is badly needed. Little use has been made of these for this book, though extensive reference to them will be found in such works as Westfall,[30] Boas,[45] Jacob[82] and Jacob.[83] Newton's papers are now more scattered and the best way to locate them is through recent work on Newton, some of which is referred to above while more may be traced through the entry on him in Gillespie.[134]

Of other scientists, Lister's papers are in the Bodleian Library, Oxford, having been transferred there from the Ashmolean Museum, to which he presented them; this was also the case with Aubrey's. For guides to these see Carr[142] and Hunter[143] respectively, in addition to the *Summary Catalogue of Western Manuscripts in the Bodleian Library at Oxford* (Oxford: Clarendon Press, 1922–53). John Evelyn's very extensive manuscript remains are at present on deposit at Christ Church, Oxford, and all quotations from them here are inserted by kind permission of the Trustees of the Will of the late J. H. C. Evelyn; the only catalogue of them currently available is a manuscript one at Christ Church. Pepys's manuscripts are partly in the Pepysian Library at Magdalene College, Cambridge, partly among the Rawlinson manuscripts in the Bodleian Library. The Bodleian also contains Thomas Smith's papers, including a considerable amount of scientific correspondence, and the Lovelace Collection of Locke manuscripts, much used in Cranston.[138]

Perhaps the most important collection in private hands are the Petty manuscripts, in the custody of the Earl of Shelburne at Bowood, Wiltshire, which are currently being catalogued under the supervision of the Bodleian. I have not seen these and hence have not used them here, but I have made extensive reference to Sharp,[144] which is largely based on them. Other manuscripts are scattered more widely, and there are many that I have not seen: for instance, there is a group of John Collins's manuscripts in the collection of the Earl of Macclesfield and various papers of John Flamsteed and Edmond Halley in the possession of the Royal Astronomical Society at Herstmonceux. Those that I have used are referred to piecemeal in the notes. Readers may like to know that a guide to groups of manuscripts of major British scientists, 1600–1940, is currently being compiled by the Royal Commission on Historical Manuscripts in association with the Royal Society. This will be published in due course, and the current drafts may be inspected at the Historical Manuscripts Commission.

Index

Bracketed numbers in italics refer to entries in the bibliographical essay, pp. 198–219.